Hands-On Science

Introduction to Biotechnology

Brian Pressley

Certified Chain of Custody
Promoting Sustainable Forest Management
www.sfiprogram.org

SGS-SFI/COC-US09/5501

1 2 3 4 5 6 7 8 9 10

ISBN 978-0-8251-6638-9

J. Weston Walch, Publisher

40 Walch Drive • Portland, ME 04103

www.walch.com

Printed in the United States of America

Contents

To the Teacher . . . . *v*

National Science Education Standards Correlations . . . . . . *vi*

General Rubric . . . . *xxiii*

1. Making Yogurt 1
2. Invasive Species 7
3. Making Kimchi 15
4. GMO Pro, GMO Con 21
5. Yeast Fermentation 29
6. Bt, Borers, and Butterflies 37
7. Antibiotics from Nature 44
8. Do You Know What You're Eating? 51
9. Gel Electrophoresis 57
10. Building People 62
11. Running a Gel 69
12. To Clone, or Not to Clone? 75
13. Was It Alive? 80
14. The Community and Genetically Modified Foods 86
15. Oil Cleanup with Bacteria 93
16. Where Will the Money Go? 100
17. Enzymes 105
18. Forensics 111
19. Extracting DNA 118
20. Biotechnology Careers 124

Glossary 129

To the Teacher

Humans were using biotechnology long before the word existed to describe what they were doing. Selective breeding was used on plants or animals with desirable traits to increase the frequency of the appearance of those traits. Recognizing that certain natural processes could be adopted, early people used fermentation, often promoted by microorganisms, to produce items such as beer, wine, cheese, yogurt, bread, sauerkraut, and kimchi. Early people also discovered that parts of many plants and animals could be used to treat illness, as naturally occurring antibiotics, and to promote healthy crop growth as fertilizers and insecticides. People using willow bark tea to treat headaches, fever, and even inflammation were unknowingly experimenting with biotechnology when they produced medicines.

All of these "classic" uses of biotechnology are still used by modern humans, but we have also added the field of modern biotechnology. This is a catchall phrase that refers to the use of living cells and organisms and their molecules, such as DNA, that we can extract or manipulate. This field encompasses work done in several diverse areas, such as bioremediation, gene therapy, genetic testing, cloning, biofuel production, waste treatment, crop production, biodegradable plastics, bioleaching, biological weapons, pharmaceutical production, the Human Genome Project, and many, many others.

The goal of this book is to address concepts and issues across many parts of biotechnological history. Students need to be aware that making cheese and extracting DNA from an onion are both forms of biotechnology. To that end, the activities are a mix of old and new. From making yogurt to performing electrophoresis, a wide swath of biotech activities will hopefully engage students' curiosity about where this field can take them in the future. As they finish each activity, you may find it helpful to have a class discussion about information that students discover, whether the activity expressly recommends it or not. The more students talk about biotechnology, the more motivated they will hopefully become.

These activities are designed to stand alone as supplements to your instruction. While the activities can be performed in any order according to your teaching plans, please note that some activities are similar in concept to others. For example, it will reinforce student understanding and save you time and effort to perform Activity 11: Running a Gel following Activity 9: Gel Electrophoresis. Also, both Activity 4: GMO Pro, GMO Con and Activity 14: The Community and Genetically Modified Foods focus on community attitudes toward and knowledge of genetically modified foods.

National Science Education Standards Correlations

The National Science Education Standards are not presented in standard outline form, so referencing a particular item is difficult without presenting the entire item. A summary of correlations is presented below. The correlations are separated into grades 5–8 and grades 9–12. The two charts allow you to look at an activity and then to check your copy of the National Science Education Standards so that you can read the specific content standard and content description.

For example, part of the correlations for the first activity, Making Yogurt, looks like this for grades 5–8:

Activity	Content standard	Bullet number	Content description	Bullet number(s)
1. Making Yogurt	A	2	Understandings about scientific inquiry	1, 4

This indicates that under Content Standard A, you will see some bulleted items. The number 2 in the third column indicates that you need to count down to the second bullet. The second bulleted item says, "Understandings about scientific inquiry." If you look under "Content Description," you will see that there is a section that has the exact same title, "Understandings about scientific inquiry." Under that section, you will find a number of bulleted items again. In this case there are seven of them, and this activity is correlated to numbers 1 and 4.

National Science Education Standards Correlations (Grades 5–8)

Activity	Content standard	Bullet number	5–8 content description	Bullet number(s)
1. Making Yogurt	A	1	Abilities necessary to do scientific inquiry	3, 4, 7
	A	2	Understandings about scientific inquiry	1, 4
	C	1	Structure and function in living systems	1, 2

(continued on next page)

Activity	Content standard	Bullet number	5–8 content description	Bullet number(s)
2. Invasive Species	A	1	Abilities necessary to do scientific inquiry	3, 4, 7
	A	2	Understandings about scientific inquiry	1, 4
	C	3	Regulation and behavior	1, 4
	E	1	Abilities of technological design	1, 2, 4
	E	2	Understandings about science and technology	4, 6
	F	5	Science and technology in society	3, 7
3. Making Kimchi	A	1	Abilities necessary to do scientific inquiry	3, 4, 7
	A	2	Understandings about scientific inquiry	1, 4
	C	1	Structure and function in living systems	1, 2
4. GMO Pro, GMO Con	A	1	Abilities necessary to do scientific inquiry	3, 4, 7
	A	2	Understandings about scientific inquiry	1, 4
	C	1	Structure and function in living systems	1, 2
	C	2	Reproduction and heredity	3, 4
	C	3	Regulation and behavior	1, 4
	C	5	Diversity and adaptations of organisms	1
	E	1	Abilities of technological design	1, 2, 4
	E	2	Understandings about science and technology	4, 6
	F	5	Science and technology in society	3, 7

(continued on next page)

Activity	Content standard	Bullet number	5–8 content description	Bullet number(s)
5. Yeast Fermentation	A	1	Abilities necessary to do scientific inquiry	3, 4, 7
	A	2	Understandings about scientific inquiry	1, 4
	C	1	Structure and function in living systems	1, 2
	C	2	Reproduction and heredity	3, 4
	E	2	Understandings about science and technology	4, 6
6. Bt, Borers, and Butterflies	A	1	Abilities necessary to do scientific inquiry	3, 4, 7
	A	2	Understandings about scientific inquiry	1, 4
	C	3	Regulation and behavior	1, 4
	C	5	Diversity and adaptations of organisms	1
	E	1	Abilities of technological design	1, 2, 4
	E	2	Understandings about science and technology	4, 6
	F	5	Science and technology in society	3, 7
7. Antibiotics from Nature	A	1	Abilities necessary to do scientific inquiry	3, 4, 7
	A	2	Understandings about scientific inquiry	1, 4
	E	2	Understandings about science and technology	4, 6
	F	5	Science and technology in society	3, 7
8. Do You Know What You're Eating?	A	1	Abilities necessary to do scientific inquiry	3, 4, 7
	A	2	Understandings about scientific inquiry	1, 4
	C	3	Regulation and behavior	1, 4
	E	2	Understandings about science and technology	4, 6
	F	5	Science and technology in society	3, 7

(continued on next page)

Activity	Content standard	Bullet number	5–8 content description	Bullet number(s)
9. Gel Electrophoresis	A	1	Abilities necessary to do scientific inquiry	3, 4, 7
	A	2	Understandings about scientific inquiry	1, 4
	C	1	Structure and function in living systems	1, 2
	C	2	Reproduction and heredity	3, 4
	F	5	Science and technology in society	3, 7
10. Building People	A	1	Abilities necessary to do scientific inquiry	3, 4, 7
	A	2	Understandings about scientific inquiry	1, 4
	C	1	Structure and function in living systems	1, 2
	C	2	Reproduction and heredity	3, 4
	C	5	Diversity and adaptations of organisms	1
	E	1	Abilities of technological design	1, 2, 4
	E	2	Understandings about science and technology	4, 6
11. Running a Gel	A	1	Abilities necessary to do scientific inquiry	3, 4, 7
	A	2	Understandings about scientific inquiry	1, 4
	E	2	Understandings about science and technology	4, 6
	F	5	Science and technology in society	3, 7

(continued on next page)

Activity	Content standard	Bullet number	5–8 content description	Bullet number(s)
12. To Clone, or Not to Clone?	A	1	Abilities necessary to do scientific inquiry	3, 4, 7
	A	2	Understandings about scientific inquiry	1, 4
	C	1	Structure and function in living systems	1, 2
	C	2	Reproduction and heredity	3, 4
	C	3	Regulation and behavior	1, 4
	C	5	Diversity and adaptations of organisms	1
	E	1	Abilities of technological design	1, 2, 4
	E	2	Understandings about science and technology	4, 6
	F	5	Science and technology in society	3, 7
13. Was It Alive?	A	1	Abilities necessary to do scientific inquiry	3, 4, 7
	A	2	Understandings about scientific inquiry	1, 4
	C	1	Structure and function in living systems	1, 2
	C	2	Reproduction and heredity	3, 4
	F	5	Science and technology in society	3, 7

(continued on next page)

Activity	Content standard	Bullet number	5–8 content description	Bullet number(s)
14. The Community and Genetically Modified Foods	A	1	Abilities necessary to do scientific inquiry	3, 4, 7
	A	2	Understandings about scientific inquiry	1, 4
	C	1	Structure and function in living systems	1, 2
	C	2	Reproduction and heredity	3, 4
	C	3	Regulation and behavior	1, 4
	C	5	Diversity and adaptations of organisms	1
	E	1	Abilities of technological design	1, 2, 4
	E	2	Understandings about science and technology	4, 6
	F	5	Science and technology in society	3, 7
15. Oil Cleanup with Bacteria	A	1	Abilities necessary to do scientific inquiry	3, 4, 7
	A	2	Understandings about scientific inquiry	1, 4
	C	1	Structure and function in living systems	1, 2
	C	2	Reproduction and heredity	3, 4
	C	3	Regulation and behavior	1, 4
	E	2	Understandings about science and technology	4, 6
	F	5	Science and technology in society	3, 7

(continued on next page)

Activity	Content standard	Bullet number	5–8 content description	Bullet number(s)
16. Where Will the Money Go?	A	1	Abilities necessary to do scientific inquiry	3, 4, 7
	A	2	Understandings about scientific inquiry	1, 4
	C	3	Regulation and behavior	1, 4
	C	5	Diversity and adaptations of organisms	1
	E	1	Abilities of technological design	1, 2, 4
	E	2	Understandings about science and technology	4, 6
	F	5	Science and technology in society	3, 7
17. Enzymes	A	1	Abilities necessary to do scientific inquiry	3, 4, 7
	A	2	Understandings about scientific inquiry	1, 4
	C	1	Structure and function in living systems	1, 2
	C	2	Reproduction and heredity	3, 4
	E	2	Understandings about science and technology	4, 6
	F	5	Science and technology in society	3, 7
18. Forensics	A	1	Abilities necessary to do scientific inquiry	3, 4, 7
	A	2	Understandings about scientific inquiry	1, 4
	C	1	Structure and function in living systems	1, 2
	C	2	Reproduction and heredity	3, 4
	E	2	Understandings about science and technology	4, 6
	F	5	Science and technology in society	3, 7

(continued on next page)

Activity	Content standard	Bullet number	5–8 content description	Bullet number(s)
19. Extracting DNA	A	1	Abilities necessary to do scientific inquiry	3, 4, 7
	A	2	Understandings about scientific inquiry	1, 4
	C	1	Structure and function in living systems	1, 2
	C	2	Reproduction and heredity	3, 4
	E	2	Understandings about science and technology	4, 6
	F	5	Science and technology in society	3, 7
20. Biotechnology Careers	A	1	Abilities necessary to do scientific inquiry	3, 4, 7
	A	2	Understandings about scientific inquiry	1, 4
	F	5	Science and technology in society	3, 7

National Science Education Standards Correlations (Grades 9–12)

Activity	Content standard	Bullet number	9–12 content description	Bullet number(s)
1. Making Yogurt	A	1	Abilities necessary to do scientific inquiry	1–6
	A	2	Understandings about scientific inquiry	1, 3, 6
	B	3	Chemical reactions	1
	E	1	Abilities of technological design	1–5
	E	2	Understandings about science and technology	1–3
	G	3	Historical perspectives	3
2. Invasive Species	A	1	Abilities necessary to do scientific inquiry	1–6
	A	2	Understandings about scientific inquiry	1, 3, 6
	C	4	Interdependence of organisms	3, 5
	E	1	Abilities of technological design	1–5
	E	2	Understandings about science and technology	1–3
	F	6	Science and technology in local, national, and global challenges	4
	G	3	Historical perspectives	3
3. Making Kimchi	A	1	Abilities necessary to do scientific inquiry	1–6
	A	2	Understandings about scientific inquiry	1, 3, 6
	B	3	Chemical reactions	1
	E	1	Abilities of technological design	1–5
	E	2	Understandings about science and technology	1–3
	G	3	Historical perspectives	3

(continued on next page)

Activity	Content standard	Bullet number	9–12 content description	Bullet number(s)
4. GMO Pro, GMO Con	A	1	Abilities necessary to do scientific inquiry	1–6
	A	2	Understandings about scientific inquiry	1, 3, 6
	C	1	The cell	1–4
	C	2	Molecular basis of heredity	1, 3
	C	4	Interdependence of organisms	3, 5
	E	1	Abilities of technological design	1–5
	E	2	Understandings about science and technology	1–3
	F	6	Science and technology in local, national, and global challenges	4
	G	3	Historical perspectives	3
5. Yeast Fermentation	A	1	Abilities necessary to do scientific inquiry	1–6
	A	2	Understandings about scientific inquiry	1, 3, 6
	B	3	Chemical reactions	1, 5
	C	1	The cell	1–4
	E	1	Abilities of technological design	1–5
	E	2	Understandings about science and technology	1–3
	G	3	Historical perspectives	3

(continued on next page)

Activity	Content standard	Bullet number	9–12 content description	Bullet number(s)
6. Bt, Borers, and Butterflies	A	1	Abilities necessary to do scientific inquiry	1–6
	A	2	Understandings about scientific inquiry	1, 3, 6
	B	3	Chemical reactions	1
	C	1	The cell	1–4
	C	2	Molecular basis of heredity	1, 3
	C	4	Interdependence of organisms	3, 5
	E	1	Abilities of technological design	1–5
	E	2	Understandings about science and technology	1–3
	F	6	Science and technology in local, national, and global challenges	4
	G	3	Historical perspectives	3
7. Antibiotics from Nature	A	1	Abilities necessary to do scientific inquiry	1–6
	A	2	Understandings about scientific inquiry	1, 3, 6
	B	3	Chemical reactions	1
	E	1	Abilities of technological design	1–5
	E	2	Understandings about science and technology	1–3
	G	3	Historical perspectives	3

(continued on next page)

Activity	Content standard	Bullet number	9–12 content description	Bullet number(s)
8. Do You Know What You're Eating?	A	1	Abilities necessary to do scientific inquiry	1–6
	A	2	Understandings about scientific inquiry	1, 3, 6
	C	1	The cell	1–4
	C	2	Molecular basis of heredity	1, 3
	E	1	Abilities of technological design	1–5
	E	2	Understandings about science and technology	1–3
	F	6	Science and technology in local, national, and global challenges	4
	G	3	Historical perspectives	3
9. Gel Electrophoresis	A	1	Abilities necessary to do scientific inquiry	1–6
	A	2	Understandings about scientific inquiry	1, 3, 6
	B	3	Chemical reactions	1
	E	1	Abilities of technological design	1–5
	E	2	Understandings about science and technology	1–3
	G	3	Historical perspectives	3
10. Building People	A	1	Abilities necessary to do scientific inquiry	1–6
	A	2	Understandings about scientific inquiry	1, 3, 6
	C	1	The cell	1–4
	C	4	Interdependence of organisms	3, 5
	E	1	Abilities of technological design	1–5
	E	2	Understandings about science and technology	1–3
	F	6	Science and technology in local, national, and global challenges	4
	G	3	Historical perspectives	3

(continued on next page)

Activity	Content standard	Bullet number	9–12 content description	Bullet number(s)
11. Running a Gel	A	1	Abilities necessary to do scientific inquiry	1–6
	A	2	Understandings about scientific inquiry	1, 3, 6
	B	3	Chemical reactions	1
	E	1	Abilities of technological design	1–5
	E	2	Understandings about science and technology	1–3
	G	3	Historical perspectives	3
12. To Clone, or Not to Clone?	A	1	Abilities necessary to do scientific inquiry	1–6
	A	2	Understandings about scientific inquiry	1, 3, 6
	C	1	The cell	1–4
	C	2	Molecular basis of heredity	1, 3
	C	4	Interdependence of organisms	3, 5
	E	1	Abilities of technological design	1–5
	E	2	Understandings about science and technology	1–3
	F	6	Science and technology in local, national, and global challenges	4
	G	3	Historical perspectives	3

Activity	Content standard	Bullet number	9–12 content description	Bullet number(s)
13. Was It Alive?	A	1	Abilities necessary to do scientific inquiry	1–6
	A	2	Understandings about scientific inquiry	1, 3, 6
	B	3	Chemical reactions	1, 5
	C	1	The cell	1–4
	C	2	Molecular basis of heredity	1, 3
	C	4	Interdependence of organisms	3, 5
	E	1	Abilities of technological design	1–5
	E	2	Understandings about science and technology	1–3
	G	3	Historical perspectives	3
14. The Community and Genetically Modified Foods	A	1	Abilities necessary to do scientific inquiry	1–6
	A	2	Understandings about scientific inquiry	1, 3, 6
	C	1	The cell	1–4
	C	2	Molecular basis of heredity	1, 3
	C	4	Interdependence of organisms	3, 5
	E	1	Abilities of technological design	1–5
	E	2	Understandings about science and technology	1–3
	F	6	Science and technology in local, national, and global challenges	4
	G	3	Historical perspectives	3

(continued on next page)

Activity	Content standard	Bullet number	9–12 content description	Bullet number(s)
15. Oil Cleanup with Bacteria	A	1	Abilities necessary to do scientific inquiry	1–6
	A	2	Understandings about scientific inquiry	1, 3, 6
	B	3	Chemical reactions	1, 5
	C	1	The cell	1–4
	C	2	Molecular basis of heredity	1, 3
	C	4	Interdependence of organisms	3, 5
	E	1	Abilities of technological design	1–5
	E	2	Understandings about science and technology	1–3
	F	6	Science and technology in local, national, and global challenges	4
	G	3	Historical perspectives	3
16. Where Will the Money Go?	A	1	Abilities necessary to do scientific inquiry	1–6
	A	2	Understandings about scientific inquiry	1, 3, 6
	C	2	Molecular basis of heredity	1, 3
	C	4	Interdependence of organisms	3, 5
	E	1	Abilities of technological design	1–5
	E	2	Understandings about science and technology	1–3
	F	6	Science and technology in local, national, and global challenges	4
	G	3	Historical perspectives	3

(continued on next page)

Activity	Content standard	Bullet number	9–12 content description	Bullet number(s)
17. Enzymes	A	1	Abilities necessary to do scientific inquiry	1–6
	A	2	Understandings about scientific inquiry	1, 3, 6
	B	3	Chemical reactions	1, 5
	C	1	The cell	1–4
	C	2	Molecular basis of heredity	1, 3
	E	1	Abilities of technological design	1–5
	E	2	Understandings about science and technology	1–3
	G	3	Historical perspectives	3
18. Forensics	A	1	Abilities necessary to do scientific inquiry	1–6
	A	2	Understandings about scientific inquiry	1, 3, 6
	B	3	Chemical reactions	1
	C	1	The cell	1–4
	E	1	Abilities of technological design	1–5
	E	2	Understandings about science and technology	1–3
	G	3	Historical perspectives	3
19. Extracting DNA	A	1	Abilities necessary to do scientific inquiry	1–6
	A	2	Understandings about scientific inquiry	1, 3, 6
	B	3	Chemical reactions	1, 5
	C	1	The cell	1–4
	C	2	Molecular basis of heredity	1, 3
	C	4	Interdependence of organisms	3, 5
	E	1	Abilities of technological design	1–5
	E	2	Understandings about science and technology	1–3
	G	3	Historical perspectives	3

(continued on next page)

Activity	Content standard	Bullet number	9–12 content description	Bullet number(s)
20. Biotechnology Careers	A	1	Abilities necessary to do scientific inquiry	1–6
	A	2	Understandings about scientific inquiry	1, 3, 6
	C	1	The cell	1–4
	E	1	Abilities of technological design	1–5
	E	2	Understandings about science and technology	1–3
	F	6	Science and technology in local, national, and global challenges	4
	G	3	Historical perspectives	3

General Rubric

Criteria	Procedures and reasoning	Strategies	Communication and use of data	Concepts and content
Level 1 **(Does not meet the standard)**	• Did not use scientific procedures or tools to collect data	• Failed to use reasoning • Failed to use a strategy • Failed to use a procedure	• Data not recorded • Conclusion based on data not reached • Could not use scientific terms, symbols, graphs, and so forth • Explanation of task not given or was not connected to data	• No use, little use, or inappropriate use of scientific terms • No use, little use, or inappropriate use of scientific theories or principles • No understanding, little understanding, or inappropriate understanding of the various properties or materials used in task
Level 2 **(Partially meets the standard)**	• Attempted to use scientific procedures or tools to collect data, but some collected data was inaccurate or incomplete	• Used reasoning, but only completed part of the task • Used a strategy, but was not effective in completing the task • Used a procedure, but could not collect data or form a conclusion	• Data recorded but not complete • Conclusion reached not fully supported by collected data • Attempted to use scientific terms, symbols, graphs, and so forth, but used incompletely and with missing components • Conclusions that were reached were not clear.	• Some use of appropriate scientific terms • Some use of appropriate scientific theories or principles • Some understanding of the various properties or materials used in the task

Criteria	Procedures and reasoning	Strategies	Communication and use of data	Concepts and content
Level 3 (Meets the standard)	• Used some scientific procedures or tools effectively to collect data with only minimal error	• Used effective reasoning • Used a strategy that allowed student to complete the task • Used a procedure, recorded data, conducted an experiment, and asked questions that could be tested	• Data recorded clearly • Conclusions essentially supported by collected data • Used scientific terms, symbols, graphs, and so forth • Conclusions that were reached were presented clearly.	• Appropriate use of scientific terms • Appropriate use of scientific theories or principles • Appropriate understanding of the various properties and materials used in the task
Level 4 (Exceeds the standard)	• Used scientific procedures and tools accurately to skillfully collect, analyze, and evaluate data	• Used advanced reasoning and connected observed effects with their causes • Designed a clear strategy, used the strategy, and adapted the strategy when necessary • Employed a procedure that took full advantage of all characteristics of the scientific method	• Data was recorded clearly and analyzed correctly. Further data was collected to clarify or to find error. • Conclusions were supported by collected data. Other questions and concepts that were suggested by the data were explored. • Scientific terms, symbols, graphs, and so forth were used. Multiple forms for presenting data were used. • The conclusions that were presented were clear. An effective presentation reviewing the task was used. All concepts were explained without need for further clarification.	• Complete and appropriate use of precise scientific terms; applied terms learned prior to activity from other activities • Complete and clear understanding of scientific theories or principles; referenced evidence in a relevant manner • Applied new knowledge to revise during process • Use of properties and materials suggested understanding beyond the scope of the activity. New questions related to material were formed.

1. Making Yogurt

TEACHER RESOURCE PAGE

INSTRUCTIONAL OBJECTIVES

Students will be able to:

- identify lactose fermentation as a tool of biotechnology
- produce yogurt at home

NATIONAL SCIENCE EDUCATION STANDARDS CORRELATIONS

GRADES 5–8

Content standard	Bullet number	Content description	Bullet number(s)
A	1	Abilities necessary to do scientific inquiry	3, 4, 7
A	2	Understandings about scientific inquiry	1, 4
C	1	Structure and function in living systems	1, 2

GRADES 9–12

Content standard	Bullet number	Content description	Bullet number(s)
A	1	Abilities necessary to do scientific inquiry	1–6
A	2	Understandings about scientific inquiry	1, 3, 6
B	3	Chemical reactions	1
E	1	Abilities of technological design	1–5
E	2	Understandings about science and technology	1–3
G	3	Historical perspectives	3

VOCABULARY

- **bacteria:** single-celled microorganisms
- **biotechnology:** the use or modification of organisms for human purposes
- **yogurt:** a dairy product that is made from milk though lactic acid fermentation

MATERIALS

- masking tape
- marker
- 600 ml milk
- spoons
- incubator
- (safety icon) hot plate or stove
- goggles
- two glass jars with covers (each able to hold at least 300 ml of milk)
- large saucepan
- food-grade thermometer
- unflavored, plain yogurt
- lab aprons
- gloves

1. Making Yogurt

TEACHER RESOURCE PAGE

HELPFUL HINTS AND DISCUSSION

Time frame: one class period
Structure: individuals/partners/groups
Location: classroom

Making yogurt from scratch is not impossible, but it is relatively expensive. Starter kits with the correct kind of bacteria can be purchased from a health food store, but generally it is easier and less expensive to use commercial yogurt as a starter. If you should purchase yogurt starter, you should follow the directions with extreme care. In all aspects of the lab, cleanliness is important. Food-grade materials should be used in the yogurt production, as students will be tasting and smelling the yogurt they produce.

Safety note: You should have the final say as to whether or not the yogurt is safe to consume before students do so. Students should not eat the material in the control jar. Students should be encouraged to be extremely careful with the hot materials and equipment used in the activity.

MEETING THE NEEDS OF DIVERSE LEARNERS

Students who need extra challenges should complete the Follow-Up Activity and the Extension Option. These students should also be encouraged to help struggling students with the use of equipment and monitoring safety concerns. Students who need extra help should be encouraged to define the vocabulary words and to keep them in a word bank for later use. They should also be allowed to work with a partner if necessary.

SCORING RUBRIC

Students meet the standard for this activity by:

- making clear observations using scientific language
- answering Concluding Questions accurately
- correctly sterilizing equipment
- properly following lab safety procedures

RECOMMENDED INTERNET SITES

- **Suite101.com—Making Yogurt at Home**
  http://healthycooking.suite101.com/article.cfm/making_yogurt_at_home
- **Tempeh info—Lactic Acid Fermentation**
  www.tempeh.info/fermentation/lactic-acid-fermentation.php

1. Making Yogurt

Teacher Resource Page

Answer Key

1. If done correctly, there should be yogurt in jar Y, and the material in jar C should not look anything like yogurt.
2. Sterilization of the jars helps ensure that the only bacteria that grows in the jar is that from the store-bought yogurt.
3. The process of making yogurt breaks down lactose, which is the material that lactose-intolerant people cannot digest.
4. The yogurt is experiencing bacterial growth while in the incubator.

Name ______________________ Date ______________________

1. Making Yogurt

Student Activity Page

Objective

To make yogurt through the fermentation of lactose

Before You Begin

Yogurt has been made in one form or another by humans for well over 1,000 years. It is one of the many early applications of **biotechnology** in which humans found an organism in nature (**bacteria** in this case) and used it to produce materials for humans to consume. Modern yogurt producers use a wide variety of bacteria, and some of them even consider the kind that they use to be a trade secret. One way yogurt is made is by turning the lactose in milk into lactic acid, which causes the curd of the yogurt to form. This formula shows the very simplified reaction of lactose to lactic acid:

$C_{12}H_{22}O_{11} + H_2O \rightarrow 4\ C_3H_6O_3$

(Lactose plus water produces lactic acid.)

Living organisms actually use many steps to produce this reaction. However, the end result is that lactic acid forms, and that is key to the process of yogurt making.

Materials

- masking tape
- marker
- 600 ml milk
- spoons
- incubator
- (safety icon) hot plate or stove
- goggles
- tablespoon
- two glass jars with covers (each able to hold at least 300 ml of milk)
- large saucepan
- food-grade thermometer
- unflavored, plain yogurt
- lab apron
- gloves

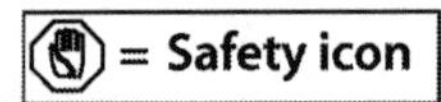

Procedure

Safety note: Use extreme caution while working with boiling water. DO NOT eat anything in the lab without the express permission of your teacher. Be sure to wear your goggles, gloves, and lab apron.

1. Put enough water in the saucepan to cover the two jars and their covers. Be sure the covers are OFF the jars so that all parts of the jars and covers are submerged and in contact with the water.
2. Bring the water to a boil.

1. Making Yogurt

3. Boil the jars and covers for 5 minutes to sterilize them.
4. Let the water cool and then remove the jars and covers to dry on clean paper towels. Do not touch the inside of the jars or covers.
5. Using a strip of masking tape and the marker, label one jar "Y" for *yogurt* and the other jar "C" for *control.*
6. Warm the milk in a saucepan to about 45°C.
7. Put half of the milk in jar Y and half in jar C.
8. Add a tablespoon of the plain yogurt to jar Y and stir well. This jar now contains the starter culture. DO NOT add yogurt to jar C.
9. Record your observations about the materials in the two jars in the "Before incubation" column of the table in the Data Collection and Analysis section.
10. Put the covers on the jars and leave them overnight in an incubator with the temperature around 45°C.
11. Remove the jars from the incubator. Record your observations in the "After incubation" column of the data table.

EXTENSION OPTION

You may *sample* some of the yogurt from jar Y, but you should not eat the material from jar C. For best results, keep the original yogurt that was used as the starter culture and refrigerate the yogurt you prepared so that they are both the same temperature. How do the tastes/smells/textures of the two yogurts compare?

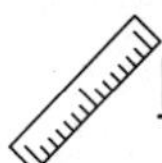

DATA COLLECTION AND ANALYSIS

Jar	Before incubation	After incubation
Y		
C		

Name ______________________ Date ______________________

1. Making Yogurt

STUDENT ACTIVITY PAGE

CONCLUDING QUESTIONS

1. What differences did you see between the material in jar Y and the material in jar C after incubation?

2. Why did you have to sterilize the jars?

3. Why are people who are lactose intolerant sometimes able to eat yogurt without any significant digestive problems?

4. What is happening to the yogurt while it sits in the incubator overnight?

FOLLOW-UP ACTIVITY

Repeat the lab, but use a wide variety of brands of plain yogurt as your starter culture. When each kind is finished, compare the taste/smell/texture of the yogurt you made with that of the starter yogurts you purchased. Record your data in a table similar to that used for this lab, with a row for each brand of yogurt you use.

2. Invasive Species

TEACHER RESOURCE PAGE

Instructional Objectives

Students will be able to:

- identify an invasive species
- determine whether the invasive species is spreading naturally or with assistance from humans
- identify a non-native species
- determine effects of the invasive species on a non-native environment

National Science Education Standards Correlations

Grades 5–8

Content standard	Bullet number	Content description	Bullet number(s)
A	1	Abilities necessary to do scientific inquiry	3, 4, 7
A	2	Understandings about scientific inquiry	1, 4
C	3	Regulation and behavior	1, 4
E	1	Abilities of technological design	1, 2, 4
E	2	Understandings about science and technology	4, 6
F	5	Science and technology in society	3, 7

Grades 9–12

Content standard	Bullet number	Content description	Bullet number(s)
A	1	Abilities necessary to do scientific inquiry	1–6
A	2	Understandings about scientific inquiry	1, 3, 6
C	4	Interdependence of organisms	3, 5
E	1	Abilities of technological design	1–5
E	2	Understandings about science and technology	1–3
F	6	Science and technology in local, national, and global challenges	4
G	3	Historical perspectives	3

Vocabulary

- **invasive species:** species that have moved into an environment where they are not typically found, generally with negative effects
- **non-native species:** species that are not normally found in a given environment

2. Invasive Species

TEACHER RESOURCE PAGE

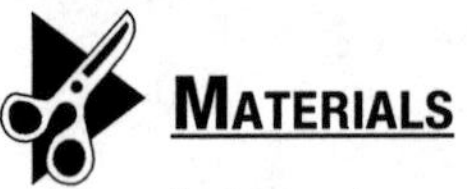

MATERIALS

- Internet access
- notebook

HELPFUL HINTS AND DISCUSSION

Time frame: one week, for research
Structure: individuals
Location: classroom, library, home

In this activity, students will use the Internet to find out more about the concepts of invasive species and non-native species, and their impact on non-native environments. Students will determine whether the species were introduced by humans or spread into a region where they had not previously been found.

MEETING THE NEEDS OF DIVERSE LEARNERS

Students who need extra challenges should complete the Follow-Up Activities and the Extension Option. These students should also be encouraged to work with struggling students to help them organize their findings with regard to the effects of invasive species moving into non-native environments. Students who need extra help may benefit from a clarification about the difference between invasive species and non-native species. They should also be allowed to work with a partner if necessary.

SCORING RUBRIC

Students meet the standard for this activity by:

- researching five different invasive species
- defining the difference between invasive species and non-native species
- determining if the species spread naturally or because of the involvement of humans
- answering the Concluding Questions
- determining some of the effects of the invasive species moving into a non-native environment
- producing a document that summarizes their findings

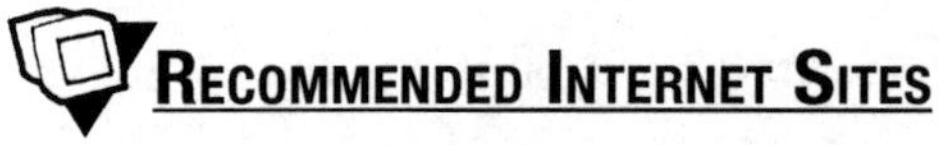

RECOMMENDED INTERNET SITES

- **Center for Invasive Species and Ecosystem Health**
  www.invasive.org
- **National Biological Information Infrastructure—Invasive Species**
  http://invasivespecies.nbii.gov/portal/server.pt
- **U.S. Fish & Wildlife Service—Invasive Species**
  www.fws.gov/invasives

2. Invasive Species

TEACHER RESOURCE PAGE

ANSWER KEY

1. Answers will vary depending on the species chosen.
2. Possible responses include: The crops will be protected; certain pests won't be able to reproduce if eaten, thereby reducing the population of the pests.
3. Possible answers include: The introduced animal might prefer a different food source, perhaps a beneficial organism; the introduced animal might overfeed on the pest and then be faced with starvation.
4. Non-native species are organisms that were not originally part of a certain environment. Invasive species are organisms that are probably non-native and have moved into an environment where they are likely to be affecting the environment in a negative manner.

Name ____________________ Date ____________________

2. Invasive Species

STUDENT ACTIVITY PAGE

OBJECTIVE

To identify **invasive species** and their effects on the local environment

BEFORE YOU BEGIN

Biotechnology is often thought of as the use of organisms by humans to improve their quality of life. This implies that humans are using biotechnology on purpose to manipulate their environment. In some cases, however, that manipulation is actually accidental. Biotechnology used by accident is still biotechnology of a sort. Take, for example, the introduction of **non-native species** to the Galapagos Islands. Sailing ships from many parts of the world have introduced animals such as pigs, rats, goats, cats, horses, cattle, donkeys, chicken, ducks, egrets, geckos, and mice. The introduction of these species was carried out by humans, and it has changed the environment. Native species have gone extinct. The new organisms are changing the way the formerly self-contained environment works.

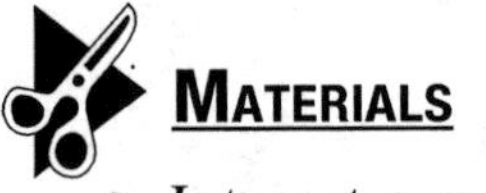

MATERIALS

- Internet access
- notebook

PROCEDURE

1. Define the term *invasive species.*
2. Define the term *non-native species.*
3. Using the Internet, find five examples of invasive species interfering with their non-native environment. For each species, determine the following and record your findings in the Invasive Species Chart in the Data Collection and Analysis section:
   - Where did it originally come from?
   - Where has it spread to?
   - Did the invasive species spread to the new environment naturally or because of human interference?
   - What effects—positive and negative—is the species having on its new environment?
4. List the address of each Web site you use in the References list in the Data Collection and Analysis section. Include the entire URL so that others may visit the site. Hand the list in to your teacher along with your Invasive Species Chart.

EXTENSION OPTION

Find some examples of humans trying to remove a non-native species from an area after humans introduced it.

Name ______________________________ Date ____________________

2. Invasive Species

STUDENT ACTIVITY PAGE

DATA COLLECTION AND ANALYSIS

INVASIVE SPECIES CHART

Invasive species	Original environment	New environment	Spread naturally or by humans?	Effects on new environment

Name ______________________ Date ______________

2. Invasive Species

STUDENT ACTIVITY PAGE

DATA COLLECTION AND ANALYSIS

REFERENCES

Species 1: ______________________ Date accessed ______________

Author name (if available) ______________________

Title of document/page ______________________

Source organization (if available) ______________________

URL ______________________

Species 2: ______________________ Date accessed ______________

Author name (if available) ______________________

Title of document/page ______________________

Source organization (if available) ______________________

URL ______________________

Species 3: ______________________ Date accessed ______________

Author name (if available) ______________________

Title of document/page ______________________

Source organization (if available) ______________________

URL ______________________

Species 4: ______________________ Date accessed ______________

Author name (if available) ______________________

Title of document/page ______________________

Source organization (if available) ______________________

URL ______________________

Species 5: ______________________ Date accessed ______________

Author name (if available) ______________________

Title of document/page ______________________

Source organization (if available) ______________________

URL ______________________

Name ______________________ Date ______________

2. Invasive Species

STUDENT ACTIVITY PAGE

CONCLUDING QUESTIONS

1. Were any of the species you researched introduced on purpose? If so, why were they introduced?

2. What benefits might come from introducing an animal to eat pests that are interfering with the growth of a certain kind of crop?

3. What drawbacks might come from introducing an animal to eat pests that are interfering with the growth of a certain kind of crop?

4. What is the difference between an invasive species and a non-native species?

Name ______________________ Date ______________

2. Invasive Species

STUDENT ACTIVITY PAGE

FOLLOW-UP ACTIVITIES

1. Raccoons have become a problem in many communities where humans have spread into the natural habitat of the raccoon or where raccoons have come into settled areas looking for food. Research to find out how communities have been dealing with raccoons as an invasive species. Present your findings to the class in a format approved by your teacher.
2. Florida has seen a rapid spread of non-native Burmese pythons, which are predatory and dangerous to humans and animals alike. Research the impact on Florida's people and environment, as well as steps taken by officials to combat the pythons. Present your findings to the class in a format approved by your teacher.

3. Making Kimchi

Teacher Resource Page

Instructional Objectives

Students will be able to:

- produce kimchi at home
- identify fermentation as a way of preserving food
- identify lactose fermentation as a tool of biotechnology

National Science Education Standards Correlations

Grades 5–8

Content standard	Bullet number	Content description	Bullet number(s)
A	1	Abilities necessary to do scientific inquiry	3, 4, 7
A	2	Understandings about scientific inquiry	1, 4
C	1	Structure and function in living systems	1, 2

Grades 9–12

Content standard	Bullet number	Content description	Bullet number(s)
A	1	Abilities necessary to do scientific inquiry	1–6
A	2	Understandings about scientific inquiry	1, 3, 6
B	3	Chemical reactions	1
E	1	Abilities of technological design	1–5
E	2	Understandings about science and technology	1–3
G	3	Historical perspectives	3

Vocabulary

- **kimchi:** a popular Korean dish that generally contains pickled cabbage
- **lactobacilli:** a kind of bacteria that usually convert lactose into lactic acid
- **microbes:** generally organisms that are microscopic in size
- **pickling:** a form of lactic acid fermentation usually used for increasing the storage time for vegetables

3. Making Kimchi

TEACHER RESOURCE PAGE

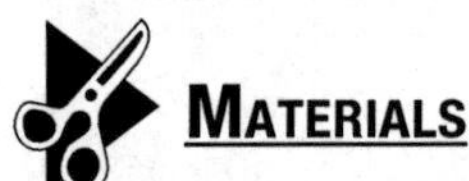

MATERIALS

- Napa cabbage or other type of Chinese cabbage
- pickling salt (non-iodized salt)
- garlic cloves
- 1 red chili pepper
- sharp knife
- pH meter
- teaspoon
- large glass jars with covers (food grade, washed with soap and water)

= Safety icon

- mixing bowl
- spoon
- goggles
- lab aprons
- gloves
- 1 cup water
- optional ingredients: a tablespoon of chopped ginger root, a teaspoon of sugar, half a cup of chopped scallions, or half a cup of chopped onions

HELPFUL HINTS AND DISCUSSION

Time frame: three to seven days
Structure: individuals or partners
Location: in class or at home

Students will need access to a clean food preparation area and will also need access to a place that is no warmer than 70°F (21°C) to store the kimchi while it is fermenting. Although there are few safety issues in an activity like this, it is still wise to have students follow normal lab safety rules such as using goggles.

MEETING THE NEEDS OF DIVERSE LEARNERS

Students who need extra challenges should complete the Follow-Up Activity and the Extension Option. These students should also be encouraged to experiment with the ratio of ingredients in later batches to establish how manipulating the variables changes the product. Students who need extra help will benefit from some instruction on safe handling of materials for food preparation.

SCORING RUBRIC

Students meet the standard for this activity by:

- making clear observations using scientific language
- answering the Concluding Questions
- correctly preparing food and containers
- correctly measuring and recording pH

3. Making Kimchi

TEACHER RESOURCE PAGE

RECOMMENDED INTERNET SITES

- **Korea Tourism Organization—Korean Cooking Classes**
  http://english.visitkorea.or.kr/enu/SI/SI_EN_3_4_8_2.jsp
- **Korean Restaurant Guide—Kimchi Histories**
  www.koreanrestaurantguide.com/kimch/kimch_0.htm
- **Love That Kimchi.com**
  www.lovethatkimchi.com

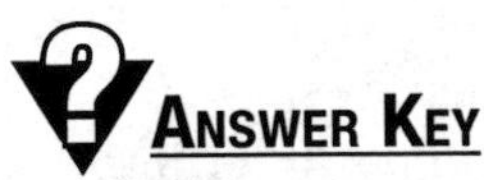

ANSWER KEY

1. Vegetables that have been pickled can go for long periods of time without spoiling. There are also a number of benefits to eating materials containing lactic acid, such as improved nutritional value of food, control of intestinal infections, improved digestion of lactose, control of some types of cancer, and control of serum cholesterol levels.
2. The cabbage mixture has undergone fermentation and as a consequence will not spoil as rapidly as untreated cabbage. It also has a different flavor, texture, and smell.
3. As the mixture continues the fermentation process, the amount of acid increases.
4. The container should not be put in a cold place because a low temperature will slow the fermentation process.

Name ______________________ Date ______________________

3. Making Kimchi

STUDENT ACTIVITY PAGE

OBJECTIVE

To make kimchi by using the pickling process of fermentation

BEFORE YOU BEGIN

Kimchi (which is also referred to as *kimchee, kim chi, gimchee,* and so forth) is a popular Korean dish made by **pickling** cabbage and other ingredients. The pickling process was originally carried out in special pots that were buried underground and dug up later. Early forms of kimchi used vegetables that Koreans could find locally and probably did not include cabbage as it does now. The pickling process allowed people to have vegetables to eat in the winter when fresh vegetables were unavailable. This early form of biotechnology uses **microbes** called **lactobacilli** to eat the natural sugars found in the cabbage and other ingredients to produce lactic acid. The amount of lactic acid eventually increases to a relatively high concentration. This helps prevent the growth of other kinds of bacteria that could cause the kimchi to spoil.

MATERIALS

- Napa cabbage or other type of Chinese cabbage
- pickling salt (non-iodized salt)
- garlic cloves
- 1 red chili pepper
- (safety icon) sharp knife
- pH meter
- teaspoon
- large glass jars with covers (food grade, washed with soap and water)
- mixing bowl
- spoon
- goggles
- lab apron
- gloves
- 1 cup water
- optional ingredients: a tablespoon of chopped ginger root, a teaspoon of sugar, half a cup of chopped scallions, or half a cup of chopped onions

(safety icon) = Safety icon

PROCEDURE

Safety note: Be extremely careful when handling the sharp knife for cutting the ingredients. Always wash your hands with soap and hot water before preparing food. After handling the chili pepper, wash your hands again and avoid touching your eyes, nose, or mouth.

1. Rinse the exterior of the cabbage under warm water.
2. Carefully chop the cabbage into inch-long pieces.
3. Thinly slice two garlic cloves.

4. Finely chop one red chili pepper. Wash your hands.
5. Combine the cabbage, garlic, and chili pepper in the mixing bowl.
6. Add three teaspoons of salt.
7. Add one cup of water.

   (*Optional:* There are many different ways to flavor kimchi. You might enjoy adding a tablespoon of chopped ginger root, a teaspoon of sugar, half a cup of chopped scallions, or half a cup of chopped onions.)
8. Stir the mixture well and put it into the large glass jar. Use more than one jar if necessary. As you're filling the jar(s), pack down the mixture with a spoon to drive out as many of the air bubbles as possible. You don't need to remove every air bubble.
9. Record some observations about the texture, taste, and smell of the mixture before the fermentation process begins. Use the table in the Data Collection and Analysis section.
10. Place the cover on each jar and leave in a cool (not cold) place for several days. Check the pH every day with the pH meter.
11. When the pH is around 3.5, the kimchi is done. (Generally this takes 3 to 7 days. If the pH is not 3.5 by the fourth or fifth day, the kimchi will still be ready to eat.)
12. If a layer of bubbly froth forms on the top of the kimchi, remove it on the last day you are going to let the kimchi ferment.
13. Now you're ready to have a bowl of homemade kimchi. As you eat, record some observations in the data table about the texture, taste, and smell of the mixture after the fermentation is finished.

Extension Option

Find some kimchi recipes on the Internet or from a cookbook. Find other versions that might better suit your taste and try making them with different ingredients.

Data Collection and Analysis

	Before fermentation	After fermentation
Texture		
Taste		
Smell		

Name ______________________ Date ______________________

3. Making Kimchi

STUDENT ACTIVITY PAGE

Concluding Questions

1. What are some of the positive benefits of pickling vegetables?

2. What effects did pickling have on the cabbage mixture?

3. The fermentation process produces lactic acid. What does the pH reaching 3.5 tell you about the nature of the fermenting mixture?

4. Why shouldn't the container be put in a cold place?

Follow-Up Activity

Using the Internet or other sources, find five other foods made using a pickling process that includes fermentation. Present your findings to the class in a format approved by your teacher.

4. GMO Pro, GMO Con

Teacher Resource Page

Instructional Objectives

Students will be able to:

- define the term *genetically modified organism (GMO)*
- determine the positive and negative outcomes of using GMOs
- determine whether people are expressing opinion, emotion, or facts when discussing GMOs

National Science Education Standards Correlations

Grades 5–8

Content standard	Bullet number	Content description	Bullet number(s)
A	1	Abilities necessary to do scientific inquiry	3, 4, 7
A	2	Understandings about scientific inquiry	1, 4
C	1	Structure and function in living systems	1, 2
C	2	Reproduction and heredity	3, 4
C	3	Regulation and behavior	1, 4
C	5	Diversity and adaptations of organisms	1
E	1	Abilities of technological design	1, 2, 4
E	2	Understandings about science and technology	4, 6
F	5	Science and technology in society	3, 7

Grades 9–12

Content standard	Bullet number	Content description	Bullet number(s)
A	1	Abilities necessary to do scientific inquiry	1–6
A	2	Understandings about scientific inquiry	1, 3, 6
C	1	The cell	1–4
C	2	Molecular basis of heredity	1, 3
C	4	Interdependence of organisms	3, 5
E	1	Abilities of technological design	1–5
E	2	Understandings about science and technology	1–3
F	6	Science and technology in local, national, and global challenges	4
G	3	Historical perspectives	3

4. GMO Pro, GMO Con

Teacher Resource Page

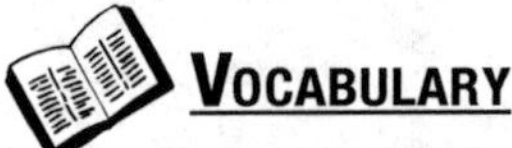

Vocabulary

- **genetic engineering:** a process of manipulating an organism's genes
- **genetically modified organism (GMO):** an organism that has had some part of its genetic makeup altered by genetic engineering

Materials

- Internet access or other resources
- notebook

Helpful Hints and Discussion

Time frame: one week for research
Structure: individuals
Location: class, library, home

Students will be using the Internet or other resources to research five specific uses of genetically modified organisms. They are to determine when and where the GMO was first used and the job it was supposed to do. They will need to make a list of all the positive outcomes of using the GMO as well as any negative side effects that have been identified. You should instruct them to try to find examples for which there are positive and negative outcomes to be discussed. Students should determine whether statements from people involved are opinions, emotion-based, or fact-based. Students will compile a bibliography of their sources, including the author name (if available), the title of the document or Web page, the source organization (if available), the URL, and the date accessed.

Meeting the Needs of Diverse Learners

Students who need extra challenges should complete the Follow-Up Activity and the Extension Option. These students should also be encouraged to present their findings in a manner other than just a simple paper, perhaps using graphs, tables, or flowcharts to illustrate their findings. Students who need extra help should be given some examples of the kinds of positive and negative outcomes that might be related to the use of GMOs to use as a guideline for selecting their own GMOs to research.

Scoring Rubric

Students meet the standard for this activity by:

- producing a document that clearly summarizes their findings
- answering the Concluding Questions
- finding five appropriate GMO examples
- collecting pertinent data on the five GMOs

4. GMO Pro, GMO Con

TEACHER RESOURCE PAGE

Recommended Internet Sites

- **GMO Compass**
  www.gmo-compass.org/eng/home
- **Human Genome Project Information—Genetically Modified Foods and Organisms**
  www.ornl.gov/sci/techresources/Human_Genome/elsi/gmfood.shtml

Answer Key

1. Students should have chosen GMOs with both pro and con issues.
2. Scientists generally address the science involved, but they may hold emotion-based opinions as well.
3. Many people opposed to the use of GMOs are bothered by the lack of long-term studies that would determine the safety of GMOs.

4–6. Student opinions will vary.

Name ______________________________ Date ______________________

4. GMO Pro, GMO Con

STUDENT ACTIVITY PAGE

OBJECTIVE

To become familiar with the nature of genetically modified organisms and the possible good and/or bad results that may come from using them

BEFORE YOU BEGIN

Genetically modified organisms (GMOs) are an important tool for people involved in biotechnology. GMOs are organisms that have had some part of their genetic makeup altered using one of the many techniques of **genetic engineering.** GMOs are used in agriculture, the production of pharmaceuticals, medical treatment, and many other places. Because GMOs are relatively new, there have been very few studies about the long-term effects of using such organisms. However, it is clear that in the present, such organisms are increasing crop yields, cleaning pollution, curing diseases, and increasing our ability to produce materials beneficial to humans.

MATERIALS

- Internet access or other resources
- notebook

PROCEDURE

1. Using the Internet, magazines, journals, newspapers, or other sources, find five examples of GMOs being used by humans to address a problem. List the source you used to learn about each example on the References page in the Data Collection and Analysis section. Provide the necessary information for each type of source, following the bibliography format preferred in your school.
2. Record the following information for each GMO in the Genetically Modified Organisms table in the Data Collection and Analysis section:
   - Name of GMO
   - Year the GMO was first used
   - Location where the GMO was first used, if available
   - Intended purpose of GMO
   - Positive outcomes
   - Negative outcomes
   - Examples of concerned groups trying to prevent the use of the GMO; these may include scientists, farmers, and/or consumers.

EXTENSION OPTION

Choose a GMO and only record the negative opinions of people who were opposed to its use. How many of these opinions seem to be based on facts and how many on emotions?

Name ____________________ Date ____________________

4. GMO Pro, GMO Con

STUDENT ACTIVITY PAGE

DATA COLLECTION AND ANALYSIS

GENETICALLY MODIFIED ORGANISMS

Name of GMO	Year/location it was first used	Intended purpose	Positive outcomes	Negative outcomes	Groups opposed	Reasons why they're opposed

Name ______________________ Date ______________________

4. GMO Pro, GMO Con

STUDENT ACTIVITY PAGE

DATA COLLECTION AND ANALYSIS

REFERENCES

Example 1: ______________________

____

____

____

____

Example 2: ______________________

____

____

____

____

Example 3: ______________________

____

____

____

____

Example 4: ______________________

____

____

____

____

Example 5: ______________________

____

____

____

____

Name ______________________ Date ______________________

4. GMO Pro, GMO Con

STUDENT ACTIVITY PAGE

CONCLUDING QUESTIONS

1. Did every GMO you researched have positive uses and potentially negative outcomes associated with it?

____

____

____

____

2. What kind of opinions did scientists generally express about the safety of using GMOs?

____

____

____

____

3. What kind of opinions did people opposed to using GMOs generally express?

____

____

____

____

4. In your opinion, what are some legitimate positive uses of GMOs?

____

____

____

____

5. In your opinion, what are some legitimate negative outcomes from using GMOs?

____

____

____

____

Name ______________________ Date ______________________

4. GMO Pro, GMO Con

STUDENT ACTIVITY PAGE

6. In your opinion, do you think the pros generally outweigh the cons, or do the cons generally outweigh the pros? Why?

____

____

____

____

Follow-Up Activity

Take an example of technology being used that does not necessarily involve biotechnology, such as computers, cell phones, cars, or televisions. Make a list of the positive and negative outcomes of using such technologies. How do the pros and cons compare to the ones you found listed about GMOs? Present your findings to the class in a format approved by your teacher.

5. Yeast Fermentation

Teacher Resource Page

Instructional Objectives

Students will be able to:

- recognize yeast fermentation as a tool of biotechnology
- determine the rate of CO_2 production from yeast fermentation under varying conditions

National Science Education Standards Correlations

Grades 5–8

Content standard	Bullet number	Content description	Bullet number(s)
A	1	Abilities necessary to do scientific inquiry	3, 4, 7
A	2	Understandings about scientific inquiry	1, 4
C	1	Structure and function in living systems	1, 2
C	2	Reproduction and heredity	3, 4
E	2	Understandings about science and technology	4, 6

Grades 9–12

Content standard	Bullet number	Content description	Bullet number(s)
A	1	Abilities necessary to do scientific inquiry	1–6
A	2	Understandings about scientific inquiry	1, 3, 6
B	3	Chemical reactions	1, 5
C	1	The cell	1–4
E	1	Abilities of technological design	1–5
E	2	Understandings about science and technology	1–3
G	3	Historical perspectives	3

Vocabulary

- **bioremediation:** the process of breaking down or removing toxins from the environment using living organisms or components of living organisms, such as plants, enzymes, fungi, and any of the many kinds of microorganisms
- **fungi:** organisms of the taxonomic kingdom Fungi that include mushrooms, molds, toadstools, and yeasts
- **yeast:** generally a unicellular organism from the Fungi kingdom

5. Yeast Fermentation

TEACHER RESOURCE PAGE

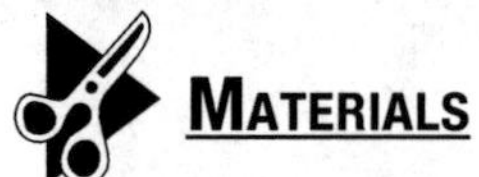

Materials

- six 125-ml Erlenmeyer flasks
- yeast
- sugar (sucrose)
- distilled water
- ice
- six balloons
- scales
- weighing papers
- six rubber bands
- tape
- thermometer
- insulating material
- string
- ruler or meterstick
- stopwatch
- goggles
- lab aprons
- gloves
- 500-ml beaker
- two 100-ml beakers
- colored pencils or markers (six different colors for graphing data)

Helpful Hints and Discussion

Time frame: 90 minutes
Structure: partners
Location: classroom

Students will keep track of CO_2 gas being produced by yeast. The lab is designed to allow them to compare two variables: the amount of sugar in the mixture, and the temperature of the mixture. Flask 1 is a control with no sugar, and flask 4 is doing double duty by having 3.0 g of sugar at 37°C. This makes it part of the 1.0 g, 2.0 g, and 3.0 g of sugar comparison, and the 5°C, room temperature, and 37°C comparisons, each with 3.0 g of sugar.

Meeting the Needs of Diverse Learners

Students who need extra challenges should complete the Follow-Up Activity and the Extension Option. These students should also be encouraged to work with struggling students with the use of equipment. Pair students who need extra help with peers who can offer support.

Scoring Rubric

Students meet the standard for this activity by:

- correctly labeling and preparing flasks
- collecting all necessary data
- producing a graph that is clear
- answering the Concluding Questions

5. Yeast Fermentation

Teacher Resource Page

Recommended Internet Sites

- **Breadworld.com—The Science of Yeast**
  www.breadworld.com/Science.aspx
- **Fine Cooking—Yeast's Crucial Roles in Breadbaking**
  www.finecooking.com/articles/yeast-role-bread-baking.aspx
- **The Story of Yeast**
  www.exploreyeast.com/story_yeast.php

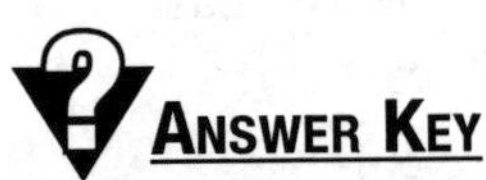

Answer Key

1. Flask 4 should have the highest amount of CO_2.
2. Adding sugar increases the amount of CO_2 formed.
3. Warmer temperatures increase the amount of CO_2 formed.
4. Based on the activity, bread should rise well if it has enough sugar in it and a warm place to rise.
5. The yeast in flask 1 did not have any nutrients to eat and produced little, if any, CO_2.

Name ______________________________ Date ______________________

5. Yeast Fermentation

STUDENT ACTIVITY PAGE

OBJECTIVE

To measure the amount of CO_2 formed by yeast under varying conditions

BEFORE YOU BEGIN

Yeast is a tool that humans have used to their advantage for thousands of years. Yeast is a living microorganism. It is a type of **fungi** that humans use in a number of areas of biotechnology. It is used to make bread, beer, in **bioremediation,** in industrial production of ethanol, as a source of B-complex vitamins, and for a number of other jobs. Modern producers of yeast search active colonies of yeast with a powerful microscope looking for particularly healthy yeast cells. Once an ideal yeast cell is found, it can be removed and will become the basis for an entire batch of yeast that may end up being tens of thousands of gallons in size. That yeast can then be used in a variety of ways, including in the making of bread, where the fermentation process makes CO_2 gas that allows the bread to rise.

MATERIALS

- six 125-ml Erlenmeyer flasks
- yeast
- sugar (sucrose)
- distilled water
- ice
- six balloons
- scales
- weighing papers
- six rubber bands
- tape
- thermometer
- insulating material
- string
- ruler or meterstick
- stopwatch
- goggles
- lab apron
- gloves
- colored pencils or markers
- 500-ml beaker
- two 100-ml beakers

PROCEDURE

Note: Timing is very important in this activity. Read these steps carefully before continuing. You will need the water of various temperatures to be ready before you begin adding it to the flasks. Do not add water to all of the flasks at once. Set your stopwatch to zero, but don't start it until directed to do so. Once you start the stopwatch, keep it running—don't reset it for each flask.

1. Prepare the following amounts of water at the given temperatures:
    - 400 ml at 37°C
    - 100 ml water at 5°C (ice water)
    - 100 ml water at room temperature

Name ______________________ Date ______________________

5. Yeast Fermentation

STUDENT ACTIVITY PAGE

Flask	Yeast	Sugar
1	3.0 g	none
2	3.0 g	1.0 g
3	3.0 g	2.0 g
4	3.0 g	3.0 g
5	3.0 g	3.0 g
6	3.0 g	3.0 g

2. Label the Erlenmeyer flasks with the numbers 1, 2, 3, 4, 5, and 6.
3. Add yeast and sugar to the flasks, in the amounts shown in the table to the right.
4. Add 100 ml of the 37°C water to flask 1.
5. Stretch the end of a balloon over the mouth of flask 1. Be sure the air is all out of the balloon. Secure the balloon with a rubber band, then tape around the rim to try to keep the flask airtight.
6. Place the flask in the insulating material and start your stopwatch.
7. Wait 1 minute, then add 100 ml of the 37°C water to flask 2. Seal it with the balloon and tape, then place it in the insulating material. Remember to keep the stopwatch running as you time each flask.
8. Wait 1 minute and add 100 ml of the 37°C water to flask 3. Seal it with the balloon and tape, then place it in the insulating material.
9. After 1 minute, add 100 ml of the 37°C water to flask 4. Seal it with the balloon and tape, then place it in the insulating material.
10. Wait 1 minute, then add the 100 ml of ice water to flask 5. Seal it with the balloon and tape, then place it in the insulating material.
11. Wait 1 minute and add the room temperature water to flask 6. Seal it with the balloon and tape, then place it in the insulating material.
12. After 20 minutes have passed since adding water to flask 1, use the string to measure around the widest circumference of the balloon on flask 1. Straighten the part of the string that went around the balloon out flat on a desk and measure it with a ruler (in millimeters). Record this number in the data table in the Data Collection and Analysis section. Keep the stopwatch running.
13. The measurement of the balloon for flask 1 should have taken about 1 minute. This means that flask 2 is now at 20 minutes. Measure the balloon the same way you did for flask 1 and record the measurement.
14. Repeat step 13 for flasks 3 through 6 and record the measurements in the data table.
15. Repeat measurements starting at the 40-, 60-, and 80-minute marks if time allows and record the measurements.
16. Plot all of the data on the graph in the Data Collection and Analysis section. Use a different colored pencil or marker for each flask.

Extension Option

Repeat this activity using different kinds of sugar (glucose, fructose, and so forth) or different brands of yeast. Compare your results.

Name ______________________ Date ______________________

5. Yeast Fermentation

STUDENT ACTIVITY PAGE

Data Collection and Analysis

Flask	Balloon circumference after 20 minutes	Balloon circumference after 40 minutes	Balloon circumference after 60 minutes	Balloon circumference after 80 minutes
1				
2				
3				
4				
5				
6				

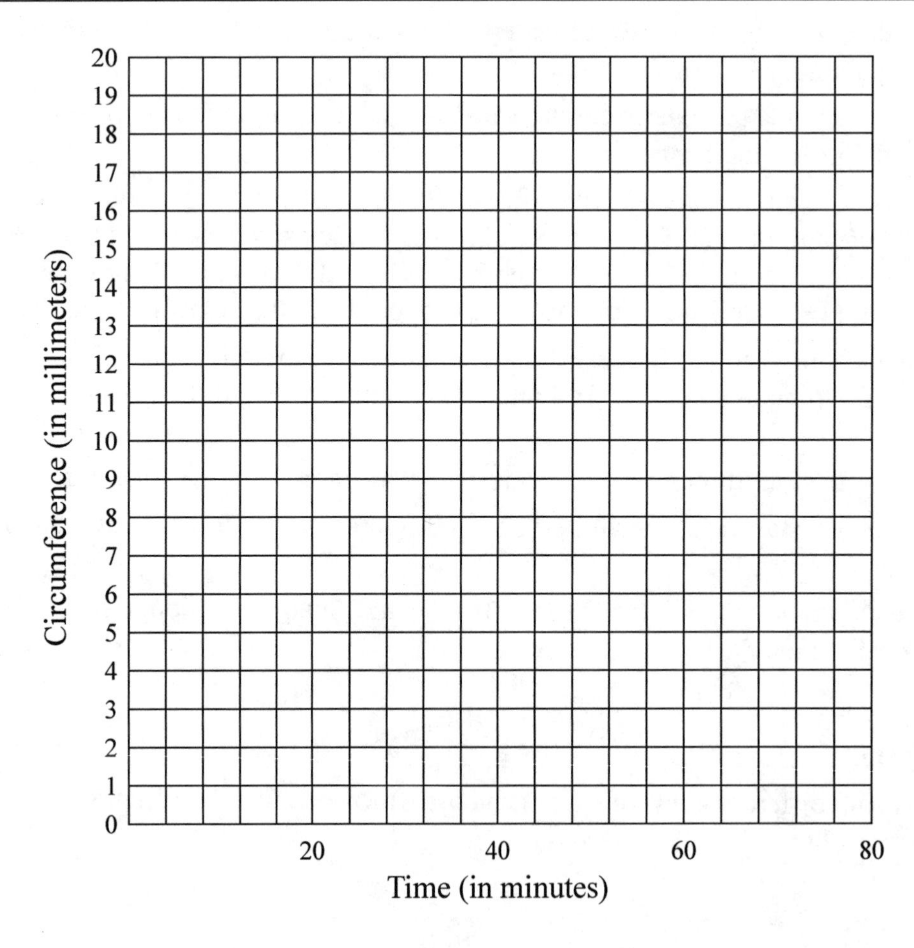

Name ____________________ Date ____________________

5. Yeast Fermentation

STUDENT ACTIVITY PAGE

CONCLUDING QUESTIONS

1. Which combination produced the largest amount of CO_2?

2. Did adding more sugar increase or decrease the amount of CO_2 produced?

3. How did the temperature of the mixture influence the production of CO_2?

4. What conditions would probably be most effective at helping bread dough rise based on your results?

Name ______________________________ Date ______________________

5. Yeast Fermentation

5. What happened to the yeast in flask 1?

____

____

____

____

____

FOLLOW-UP ACTIVITY

Research how yeast is produced and find five examples of places where yeast is used other than in bread making. Present your findings to the class in a format approved by your teacher.

6. Bt, Borers, and Butterflies

TEACHER RESOURCE PAGE

INSTRUCTIONAL OBJECTIVES

Students will be able to:

- understand the various possible outcomes from releasing genetically modified corn into the environment
- explain how science is a self-correcting discipline

NATIONAL SCIENCE EDUCATION STANDARDS CORRELATIONS

GRADES 5–8

Content standard	Bullet number	Content description	Bullet number(s)
A	1	Abilities necessary to do scientific inquiry	3, 4, 7
A	2	Understandings about scientific inquiry	1, 4
C	3	Regulation and behavior	1, 4
C	5	Diversity and adaptations of organisms	1
E	1	Abilities of technological design	1, 2, 4
E	2	Understandings about science and technology	4, 6
F	5	Science and technology in society	3, 7

GRADES 9–12

Content standard	Bullet number	Content description	Bullet number(s)
A	1	Abilities necessary to do scientific inquiry	1–6
A	2	Understandings about scientific inquiry	1, 3, 6
B	3	Chemical reactions	1
C	1	The cell	1–4
C	2	Molecular basis of heredity	1, 3
C	4	Interdependence of organisms	3, 5
E	1	Abilities of technological design	1–5
E	2	Understandings about science and technology	1–3
F	6	Science and technology in local, national, and global challenges	4
G	3	Historical perspectives	3

6. Bt, Borers, and Butterflies

VOCABULARY

- **bacillus thuringiensis (Bt):** a kind of bacteria found in soil that can be used as a pesticide
- **insecticide:** a toxin used to kill insects
- **resistant:** a characteristic of an organism that, generally through selection, has become immune to the effects of a substance
- **toxins:** poisons produced by or made from living organisms

MATERIALS

- Internet access or other resources
- notebook

HELPFUL HINTS AND DISCUSSION

Time frame: one week for research
Structure: individuals or partners
Location: class, library, home

Bt corn was first available to farmers around 1995. By 1999, the journal *Nature* published a claim that the pollen of this corn might be harming the caterpillars of monarch butterflies. Many uninformed opinions were expressed in the media and demands were made to remove Bt corn from the market. All of this was done without enough data to support the continuing use of Bt corn and without enough data to prove it was doing the harm claimed. This activity will have students build their own timeline that shows the order in which events unfolded between the time Bt corn was introduced and the present. One goal of the activity is to show students how scientists respond to claims by collecting data instead of only speculating, as the media did in this case.

Since students will compile a bibliography of their sources, consult a language arts teacher as to which format is generally used in your school, and have bibliographic reference materials handy so that students can look at examples of how to list each type of source they use.

MEETING THE NEEDS OF DIVERSE LEARNERS

Students who need extra challenges should complete the Follow-Up Activity and the Extension Option. These students should also be encouraged to work with struggling students to help them determine the salient points that should be included in their timelines. Offer students a variety of formats in which to compile their timeline, e.g., on poster boards, in PowerPoint, in teams performing a sketch, and so forth. Students who need extra help should be encouraged to work with a partner if necessary. They will also find benefit from seeing some examples of the media publishing claims without facts to back them up.

6. Bt, Borers, and Butterflies

TEACHER RESOURCE PAGE

SCORING RUBRIC

Students meet the standard for this activity by:

- answering the Concluding Questions
- determining which events were driven by opinion and which by research and facts
- making a timeline correctly showing the order of events surrounding the introduction of Bt corn

RECOMMENDED INTERNET SITES

- **United States Department of Agriculture—Butterflies and Bt Corn: Allowing Science to Guide Decisions**
  www.ars.usda.gov/sites/monarch
- **United States Department of Agriculture—Q&A: Bt Corn and Monarch Butterflies**
  www.ars.usda.gov/is/br/btcorn/index.html#bt1
- **University of Kentucky College of Agriculture—Bt-corn for Corn Borer Control**
  www.ca.uky.edu/entomology/entfacts/ef118.asp

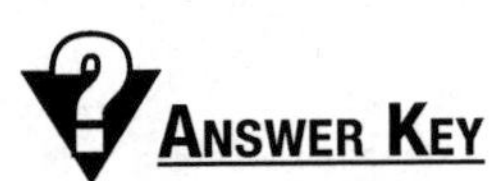

ANSWER KEY

1. Some of the benefits are increased crop yields and the ability to grow corn in regions where it was not previously possible. Students will find many others.
2. Some possible drawbacks include toxins killing non-pest organisms and increasing resistance to Bt. Students will find many others.
3. The biggest misconception was that monarch butterfly populations would be harmed by the Bt. Students may find others.
4. The biggest misconception, that monarch butterfly populations would be harmed, was shown to be unlikely by a series of experiments done by scientists in many different places where Bt corn was being raised.

Name ______________________ Date ______________________

6. Bt, Borers, and Butterflies

STUDENT ACTIVITY PAGE

OBJECTIVE

To understand the effects of releasing transgenic corn into the environment

BEFORE YOU BEGIN

Bacillus thuringiensis (**Bt**) is a bacterium commonly found in soil. It can be used as an **insecticide.** Some varieties of corn have been genetically altered to make the **toxins** naturally produced by Bt. This is so that animals, such as the corn borer, will encounter it when they eat parts of the corn plant. Using biotechnology to produce the Bt toxins kills the borer and some other kinds of pests that eat the corn. After Bt corn was being raised in a number of places, concerns were raised. Some said that the genetically modified corn was making genetically altered pollen that could poison non-pest organisms such as the monarch butterfly. Another concern was that the genetic material from the corn might spread into other plants, increasing the amount of Bt toxins in areas near where the corn was being grown. This might mean that if animals in a large area were exposed to varying levels of Bt toxins, they would eventually become **resistant** to it. Then the toxins would no longer be useful as a pesticide.

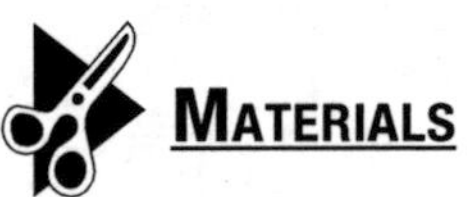

MATERIALS

- Internet access or other resources
- notebook

PROCEDURE

1. Find sources on the Internet or in print form that detail the various events that took place from the introduction of Bt corn to the present, or as close to the present as possible. Pay special attention to how scientists responded to the claims (based on very little evidence) that Bt corn was dangerous. Compile a bibliography of your sources on the References form in the Data Collection and Analysis section. Follow the bibliography format preferred at your school.
2. Collect as much information as you can about the various groups involved in the story.
3. Summarize your findings on a timeline. Discuss the format of your timeline with your teacher. Your timeline should have: dates, quotes, images, and brief summaries of events. Use the template in the Data Collection and Analysis section as a way to organize your notes before creating your final timeline.
4. Turn in your timeline and list of references to your teacher.

EXTENSION OPTION

Find another biotechnology situation similar to the introduction of Bt corn and construct a timeline for the events. Compare the methods used by the people who were for or against the use of biotechnology in this situation.

Name ____________________ Date ____________________

6. Bt, Borers, and Butterflies

Student Activity Page

Data Collection and Analysis

Genetically Modified Organism Timeline Planner

Use this sheet as a rough draft for planning which events you will include in your final timeline. Use the larger lines to portray major events, the beginning of a new 5- or 10-year period, and so forth, depending on your research.

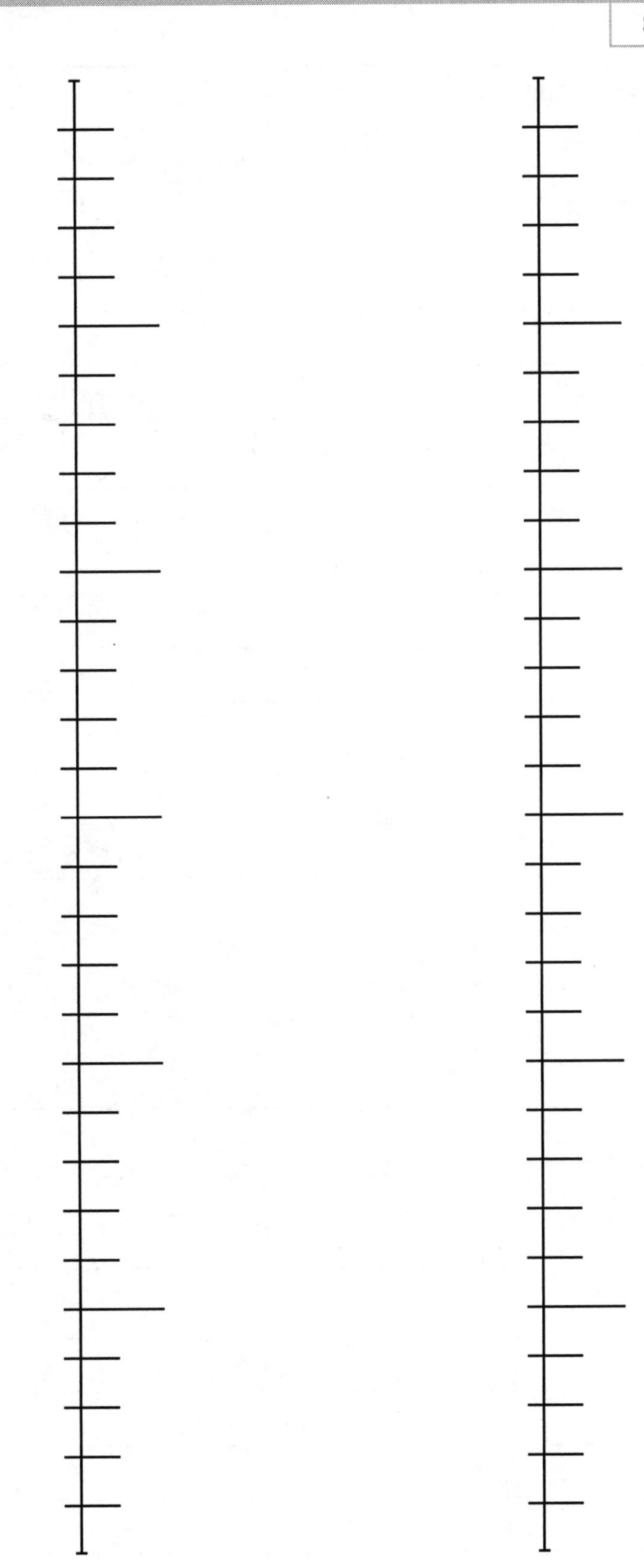

Name ______________________________ Date ______________________

6. Bt, Borers, and Butterflies

Student Activity Page

Data Collection and Analysis

References

Source 1: ______________________________

____

____

____

____

Source 2: ______________________________

____

____

____

____

Source 3: ______________________________

____

____

____

____

Source 4: ______________________________

____

____

____

____

Source 5: ______________________________

____

____

____

____

Name ______________________ Date ______________________

6. Bt, Borers, and Butterflies

STUDENT ACTIVITY PAGE

CONCLUDING QUESTIONS

1. Make a list of the possible benefits of using Bt corn.

2. Make a list of the possible drawbacks of using Bt corn.

3. What misconceptions about Bt corn appear early in your timeline?

4. How were misconceptions about Bt corn addressed by the end of your timeline?

FOLLOW-UP ACTIVITY

Find out what percent of corn grown worldwide is genetically modified. What other crops are usually grown in nearby fields? Why? Present your findings to the class in a format approved by your teacher.

7. Antibiotics from Nature

TEACHER RESOURCE PAGE

INSTRUCTIONAL OBJECTIVES

Students will be able to:

- determine which naturally occurring materials have promise as antibiotics
- compare the effectiveness of natural antibiotics to an over-the-counter antibiotic
- connect the production of medicine to fields involved in biotechnology

NATIONAL SCIENCE EDUCATION STANDARDS CORRELATIONS

GRADES 5–8

Content standard	Bullet number	Content description	Bullet number(s)
A	1	Abilities necessary to do scientific inquiry	3, 4, 7
A	2	Understandings about scientific inquiry	1, 4
E	2	Understandings about science and technology	4, 6
F	5	Science and technology in society	3, 7

GRADES 9–12

Content standard	Bullet number	Content description	Bullet number(s)
A	1	Abilities necessary to do scientific inquiry	1–6
A	2	Understandings about scientific inquiry	1, 3, 6
B	3	Chemical reactions	1
E	1	Abilities of technological design	1–5
E	2	Understandings about science and technology	1–3
G	3	Historical perspectives	3

VOCABULARY

- **antibiotic:** a substance that kills bacteria or at least hinders the growth of bacteria

7. Antibiotics from Nature

TEACHER RESOURCE PAGE

MATERIALS

- sterile filter paper
- sterile tweezers or sterile forceps
- over-the-counter antibacterial ointment
- potential natural antibiotics:
  - garlic
  - salt
  - aloe vera
  - green tea extract
  - honey
  - tea tree extract
  - lemon juice, etc.
- nutrient agar plates of *E. coli*
- distilled water
- sterile hole punch
- gloves
- goggles
- sterile 50-ml beakers or other similar small containers
- glass stirring rod
- incubator
- ruler
- lab aprons

HELPFUL HINTS AND DISCUSSION

Time frame: two days
Structure: individuals or partners
Location: classroom

Students will test a variety of naturally occurring substances to see if they have any promise as antibiotic agents. They will also use an over-the-counter antibiotic as a comparison. They will need nutrient agar plates that have been inoculated with *E. coli.* Although *E. coli* is generally safe for humans to handle, it would be best if you prepared the plates of *E. coli.* If you are unfamiliar with the procedure, check the second site listed under Recommended Internet Sites for an overview. The list of potential natural antibiotics to test is a suggested list, and students may bring in other substances to test if they wish. Students will be putting four samples on each plate, so you will need a large number of plates. You might consider having students limit their number of samples to eight or twelve, including the over-the-counter antibacterial ointment and one disc of sterile distilled water to use as a control. To clean all of the materials that are contaminated with *E. coli,* soak them in a 10% solution of bleach for at least 2 hours.

MEETING THE NEEDS OF DIVERSE LEARNERS

Students who need extra challenges should complete the Follow-Up Activity and the Extension Option. These students should also be encouraged to work with struggling students with the use of equipment and maintaining a sterile work environment. Students who need extra help will benefit from instruction on how the agar surface will appear with and without growth. They should also be allowed to work with a partner if necessary.

7. Antibiotics from Nature

TEACHER RESOURCE PAGE

SCORING RUBRIC

Students meet the standard for this activity by:

- carefully handling all lab materials
- recording accurate information in the data table
- correctly following safety requirements
- answering the Concluding Questions

RECOMMENDED INTERNET SITES

- **Centers for Disease Control and Prevention—*Escherichia coli***
  www.cdc.gov/nczved/dfbmd/disease_listing/stec_gi.html
- **Science Buddies—Which Acne Medication Can Really Zap That Zit? (provides a description of how to prepare agar plates)**
  www.sciencebuddies.org/science-fair-projects/project_ideas/MicroBio_p019.shtml

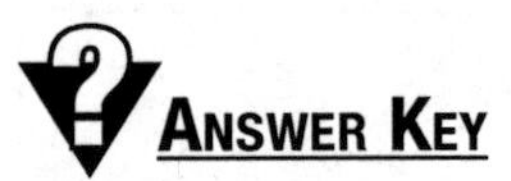

ANSWER KEY

1–3. Answers will vary depending on the substances chosen.

4. The antibacterial ointment should show superior ability as an antibiotic.
5. The growth should have been unaffected by the distilled water.

Name ______________________________ Date ______________________

7. Antibiotics from Nature

STUDENT ACTIVITY PAGE

OBJECTIVE

To determine the effectiveness of a number of naturally occurring substances as antibiotics

BEFORE YOU BEGIN

Antibiotics are substances that kill bacteria or at least hinder the growth of bacteria. Humans have become good at making new antibiotics in a number of ways. The main way, however, still starts with living organisms. Researchers observe a wide variety of organisms, such as plants, animals, and microorganisms, and find out how they interact with organisms in their environment. If they make substances that slow the growth of bacteria or kill bacteria, then researchers may study them further to see if the substances can be used by humans. This is one of the ways that biotechnology is used to make medicine. This activity will give you the chance to explore the natural antibiotic properties of some common substances.

MATERIALS

- sterile filter paper
- sterile tweezers or sterile forceps
- over-the-counter antibacterial ointment
- potential natural antibiotics:
  - garlic
  - salt
  - aloe vera
  - green tea extract
  - honey
  - tea tree extract
  - lemon juice, etc.
- (Safety icon) nutrient agar plates of *E. coli*
- distilled water
- sterile hole punch
- gloves
- goggles
- sterile 50-ml beakers or other similar small containers
- glass stirring rod
- incubator
- ruler
- lab apron

(Safety icon) = Safety icon

PROCEDURE

1. Be sure that all of the work surfaces are as clean as possible.
2. Put on your gloves and goggles and lab apron.
3. Use the sterile hole punch to punch out a number of small discs from a piece of filter paper. Collect them in one of the sterile 50-ml beakers.
4. Prepare your potential antibiotics. Place a small amount of each potential antibiotic in its own 50-ml beaker. Place sterile distilled water in one of the beakers to act as a control.

7. Antibiotics from Nature

5. If the material is not liquid, add 10 ml of distilled water to the beaker and then mash the material as much as possible with the glass stirring rod. Be sure to clean the rod between substances or to use a new rod for each substance.
6. Add one small filter paper disc to each beaker. Be sure the disc is covered with the liquid from the material, but avoid getting any large pieces of the samples on the disc.
7. Determine how many *E. coli* plates you will need. You will put four discs on each *E. coli* plate. In addition to the substances in your beakers, you will also test an over-the-counter antibiotic ointment, and you will include one disc of sterile distilled water to use as a control. Collect enough plates from your teacher for all of the substances you will be testing.
8. Turn each plate upside down (so that the agar is at the top) and draw lines on the outside of the plate to divide it into quadrants as shown in the diagram. Number the quadrants 1 through 4 on the first plate, 5 through 8 on the next plate, and so forth.
9. Using the forceps, collect the small paper disc from your first potential antibiotic and place it carefully in the center of the agar in quadrant 1. In your data table in the Data Collection and Analysis section, record this as substance 1.

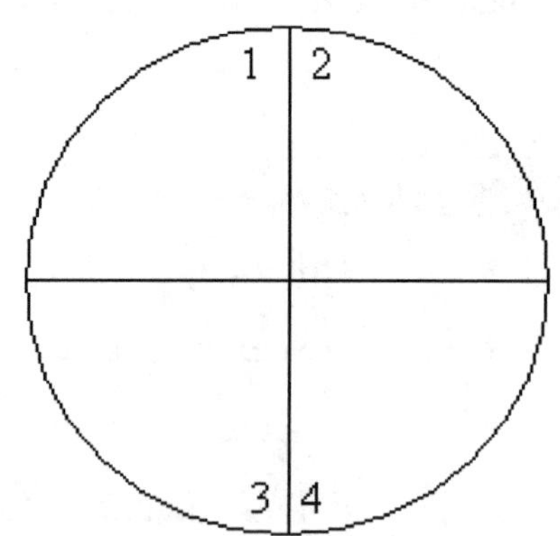

10. Clean your forceps and repeat step 9 for each of your potential substances. Be sure to record the number associated with each substance in your data table. If you need more room, create a similar data table on your own paper.
11. If you have not done so already, be sure to include one paper disc with the over-the-counter antibiotic ointment in one of your plates. Also, include one paper disc with sterile distilled water to use as a control.
12. Put the cover back on your plate and place it in the incubator at 37°C for at least 48 hours. Be sure to invert the plate so that the cover is on the bottom and the agar and samples are on the "ceiling" of the container.
13. If there is no clear growth on the plate after 48 hours, put it back in the incubator until growth is seen.
14. Remove the samples from the incubator and DO NOT open them. Using a ruler, measure the diameter of the area around each substance (in mm) where there is no *E. coli* growth. Record the measurements in the data table. If the sample is covered in growth, record the diameter as zero.
15. Return your samples to the teacher for disposal. DO NOT open them.

EXTENSION OPTION

Repeat this activity using a number of known antibiotic ointments to compare their effectiveness.

Name ______________________________ Date ______________________________

7. Antibiotics from Nature

STUDENT ACTIVITY PAGE

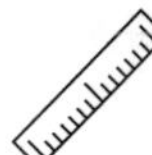

Data Collection and Analysis

Substance	Diameter (mm)	Substance	Diameter (mm)
1		7	
2		8	
3		9	
4		10	
5		11	
6		12	

Concluding Questions

1. Which of the natural samples showed some promise as an antibiotic?

____

____

____

____

2. Which natural sample provided the widest diameter with no growth?

____

____

____

____

3. Which samples showed little or no promise as an antibiotic?

____

____

____

____

Name ______________________________ Date ______________________

7. Antibiotics from Nature

STUDENT ACTIVITY PAGE

4. How did the antibacterial ointment perform compared to the natural substances?

____

____

____

____

5. What was the growth like around your sterile distilled water control disc?

____

____

____

____

FOLLOW-UP ACTIVITY

Find out the name of a widely used antibiotic that is derived from an organism. What is the organism and what substance does it produce that is used as an antibiotic? Present your findings to the class in a format approved by your teacher.

8. Do You Know What You're Eating?

TEACHER RESOURCE PAGE

INSTRUCTIONAL OBJECTIVES

Students will be able to:

- determine the amount of genetically modified food produced in the United States
- determine people's perception about the amount of genetically modified food produced in the United States
- understand the various opinions people hold about the use of genetically modified foods produced in the United States

NATIONAL SCIENCE EDUCATION STANDARDS CORRELATIONS

GRADES 5–8

Content standard	Bullet number	Content description	Bullet number(s)
A	1	Abilities necessary to do scientific inquiry	3, 4, 7
A	2	Understandings about scientific inquiry	1, 4
C	3	Regulation and behavior	1, 4
E	2	Understandings about science and technology	4, 6
F	5	Science and technology in society	3, 7

GRADES 9–12

Content standard	Bullet number	Content description	Bullet number(s)
A	1	Abilities necessary to do scientific inquiry	1–6
A	2	Understandings about scientific inquiry	1, 3, 6
C	1	The cell	1–4
C	2	Molecular basis of heredity	1, 3
E	1	Abilities of technological design	1–5
E	2	Understandings about science and technology	1–3
F	6	Science and technology in local, national, and global challenges	4
G	3	Historical perspectives	3

VOCABULARY

- **by-products:** secondary products produced during or after the production of a main product, such as eggs from chickens, assuming chickens are the main product
- **genetically modified food:** food from an organism that has had some part of its genetic makeup altered by genetic engineering

8. Do You Know What You're Eating?

TEACHER RESOURCE PAGE

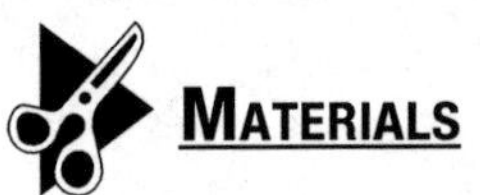

MATERIALS

- Internet access or other resources
- notebook

HELPFUL HINTS AND DISCUSSION

Time frame: one week for research, plus one or more class periods for student presentations
Structure: individuals or partners
Location: outside of school, in class

A large part of the crops grown in the United States have been genetically modified in some way, at least for some specific crops such as soy and corn. Many of the by-products of these crops are in foods that most consumers purchase and eat, but consumers remain largely ignorant of that fact. This activity will help students raise their awareness of how prevalent GM foods are becoming, and of the perceptions of the public regarding that issue. Students will find that the general population knows very little about GM foods and whether they are safe. Students should try to survey in excess of 20 people, and the more people they can survey the better. All students will benefit from a discussion of how to write a survey that is as free from bias as possible. Review students' survey questions for bias. Allow sufficient class time for students to present their findings.

MEETING THE NEEDS OF DIVERSE LEARNERS

Students who need extra challenges should complete the Follow-Up Activity and the Extension Option. Students who need extra help should be assisted in finding Web sites or other references and should have some help in constructing their survey in an unbiased manner. They should also be allowed to work with a partner if necessary.

SCORING RUBRIC

Students meet the standard for this activity by:

- collecting data on current GM food usage
- preparing an unbiased survey
- administering the survey
- collecting and presenting the data of the survey in a clear manner
- answering the Concluding Questions

8. Do You Know What You're Eating?

TEACHER RESOURCE PAGE

Recommended Internet Sites

- **Institute for Responsible Technology—Genetically Modified Ingredients Overview**
  www.responsibletechnology.org/GMFree/AboutGMFoods/GMFoodsataGlance/index.cfm
- **Institute for Responsible Technology—The Basics: What's a GMO**?
  www.responsibletechnology.org/GMFree/AboutGMFoods/FAQs/index.cfm#what_is_a_gmo
- **WebMD—Are Biotech Foods Safe to Eat?**
  www.webmd.com/diet/features/are-biotech-foods-safe-to-eat
- **World Health Organization—20 Questions on Genetically Modified (GM) Foods**
  www.who.int/foodsafety/publications/biotech/20questions/en/

Answer Key

1. Students will generally find that canola, corn, cotton, and soy are major crops with a large percentage of GM products.

2–5. Answers will vary depending on survey results.

Name ________________________________ Date ________________________

8. Do You Know What You're Eating?

STUDENT ACTIVITY PAGE

OBJECTIVES

- To become familiar with the amount of genetically modified (GM) food that is produced in the United States
- To conduct an unbiased survey of people's perceptions about genetically modified food

BEFORE YOU BEGIN

Biotechnology has made a wide variety of **genetically modified foods** available in the United States. Crops such as canola, corn, cotton, and soy are widely produced from genetically modified stock. Many animals that humans consume, such as pigs, cows, and chickens, and their **by-products,** such as milk, cheese, and eggs, may be affected. This is because many of these animals are given feed that contains genetically modified materials. Some sources say that more than 80 percent of the soybeans raised in the United States come from genetically modified stock. This means that the beans themselves and all of their by-products are potentially "contaminated" because they are not "all natural."

MATERIALS

- Internet access or other resources
- notebook

PROCEDURE

1. Using the Internet or published resources such as books, magazines, newspapers, journals, and so forth, find as much information as you can on the amount of genetically modified food that is raised in the United States. Record the kind of food, the amount that is genetically modified, and the percentage of the total crop that GM food represents in the data table in the Data Collection and Analysis section. If you need more space, create a similar data table on your own paper.
2. Create a survey that tests people's knowledge about the amount of GM foods they purchase or eat. Your survey should include such questions as:
   - Do you know what a GM food is?
   - Do you purchase GM foods?
   - If a product had a label indicating it was a GM food, would you still purchase it?
   - Do you think GM foods are safe?
   - Do you know if the government has a department that approves GM foods for human consumption?
   - How much GM food do you think is produced in the United States?
   - Include several questions of your own to include in your survey. Make sure that the questions do not hint at your personal opinion about GM foods. Unbiased questions are vital to gathering accurate data.

8. Do You Know What You're Eating?

STUDENT ACTIVITY PAGE

3. Review your questions with your teacher before conducting your survey. Your teacher will determine the number of people whom you will survey. Try to get a wide variety of people to respond to your survey; that is, don't just survey other students.
4. One good place to give your survey might be in front of a local supermarket with the permission of the manager. Be sure to show them the survey and stress that you are not presenting information that is for or against buying GM foods, but are instead trying to find out what people know about them. Make sure the manager knows that this is for a school project.
5. Make a series of graphs or a poster that summarizes your findings. Present your findings to the class.

Extension Option

Conduct a follow-up safety survey that only asks questions about the general safety of food products. Based on the results of the second survey, how do you think people would have responded to your first survey if they were told that all GM foods were unsafe before they took the survey? What if they were told that all GM foods are perfectly safe? Present your findings to the class.

Data Collection and Analysis

Type of food	Amount that is genetically modified	Percentage of total crop

Concluding Questions

1. What crops have a large percentage of genetically modified components?

____

____

____

____

Name ____________________ Date ____________________

8. Do You Know What You're Eating?

STUDENT ACTIVITY PAGE

2. What percentage of people were unaware they were probably purchasing GM foods?

3. What percentage of people seemed well-informed about the presence of GM foods in the United States?

4. Now that you have collected and presented your data, what survey questions, if any, would you change and how would you change them?

5. What questions would you add to or remove from your survey if you were doing it over?

FOLLOW-UP ACTIVITY

Write a survey about the uses of biotechnology. Use the survey to find out if people view biotechnology as positive or negative. Present your findings to the class.

9. Gel Electrophoresis

Teacher Resource Page

Instructional Objectives

Students will be able to:

- understand electrophoresis as a tool of biotechnology
- understand the basic method by which electrophoresis works

National Science Education Standards Correlations

Grades 5–8

Content standard	Bullet number	Content description	Bullet number(s)
A	1	Abilities necessary to do scientific inquiry	3, 4, 7
A	2	Understandings about scientific inquiry	1, 4
C	1	Structure and function in living systems	1, 2
C	2	Reproduction and heredity	3, 4
F	5	Science and technology in society	3, 7

Grades 9–12

Content standard	Bullet number	Content description	Bullet number(s)
A	1	Abilities necessary to do scientific inquiry	1–6
A	2	Understandings about scientific inquiry	1, 3, 6
B	3	Chemical reactions	1
E	1	Abilities of technological design	1–5
E	2	Understandings about science and technology	1–3
G	3	Historical perspectives	3

Vocabulary

- **electrophoresis:** the motion of electrically charged particles in a material that is permeated by an electric field

Materials

- electrophoresis unit (safety icon)
- power supplies (safety icon)
- graduated cylinder
- transfer pipettes
- tap water
- bromothymol blue solution (0.5%)
- glass stirring rod
- goggles
- lab aprons
- gloves

9. Gel Electrophoresis

TEACHER RESOURCE PAGE

HELPFUL HINTS AND DISCUSSION

Time frame: 30 minutes

Structure: individuals, partners, or groups

Location: classroom

Students will run an electrophoresis chamber with only water and an indicator in it. The process is essentially electrolysis, and shows students how negatively charged particles are drawn to the anode and positively charged particles are drawn to the cathode. In this lab, as H_2 gas is released at the cathode, OH^- is left behind, which makes the water basic around the cathode. The water there should turn dark blue. As O_2 gas is released at the anode, H_3O^+ is left behind, which makes the water acidic around the anode. The water there should turn yellow. As always, students should be extremely careful when working with water and electricity.

MEETING THE NEEDS OF DIVERSE LEARNERS

Students who need extra challenges should complete the Follow-Up Activity and the Extension Option. These students should also be encouraged to use the correct scientific terminology (such as *cathode* and *anode*). Students who need extra help will benefit from a step-by-step walk-through of the procedure before the lab.

SCORING RUBRIC

Students meet the standard for this activity by:

- setting up equipment, making observations, and recording them in the data table
- correctly explaining the concept of electrophoresis

RECOMMENDED INTERNET SITES

- **Colorado State University—Agarose Gel Electrophoresis of DNA**
  www.vivo.colostate.edu/hbooks/genetics/biotech/gels/agardna.html
- **HyperPhysics Thermodynamics—Electrolysis of Water**
  http://hyperphysics.phy-astr.gsu.edu/Hbase/thermo/electrol.html
- **The University of Tennessee at Knoxville—Gel Electrophoresis**
  http://web.utk.edu/~khughes/GEL/sld001.htm

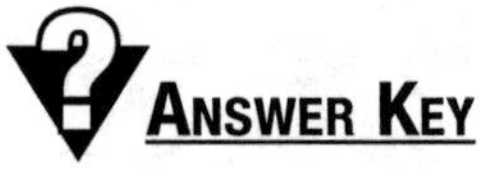

ANSWER KEY

1. The water should turn yellow, indicating the presence of an acid.
2. The water should turn dark blue, indicating the presence of a base.
3. The positively charged molecules are attracted to the negative terminal.
4. The negatively charged molecules are attracted to the positive terminal.

Name ______________________________ Date ______________________

9. Gel Electrophoresis

STUDENT ACTIVITY PAGE

OBJECTIVE

To understand the mechanism that allows electrophoresis to work

BEFORE YOU BEGIN

Gel **electrophoresis** is a process that allows molecules such as DNA to be separated into various smaller molecules. The final product of electrophoresis allows the user to determine a wide variety of properties related to the molecule that is being tested. Gel electrophoresis can be used to make a DNA profile of a person. This profile can be used to identify genetic diseases, to identify a criminal, or to establish paternity. Such tests are used in many fields, and biotechnology is one of those fields. Researchers use these tests to search the DNA of many organisms for traits that they can study further. Their work can then be applied to medicine, agriculture, embryology, botany, biology, and many other scientific fields.

MATERIALS

- electrophoresis unit (safety icon)
- power supplies (safety icon)
- graduated cylinder
- transfer pipettes
- tap water
- bromothymol blue solution (0.5%)
- glass stirring rod
- goggles
- lab apron
- gloves

PROCEDURE

1. Set up your electrophoresis chamber and power supply as directed, but do not plug in the supply. You may leave the gel tray out of the chamber, since it is not necessary for this activity.
2. Fill a large beaker with tap water.
3. Pour water into the electrophoresis chamber until it is full enough to make sure that the electrodes are in the water deep enough for a current to run.
4. Add 0.5 ml of the 0.5% bromothymol blue solution.
5. Stir the water with the glass stirring rod to ensure that the indicator is evenly spread throughout the chamber.
6. Record your observations about the color of the water in the chamber in the data table in the Data Collection and Analysis section.
7. Carefully inspect the anode and cathode before you turn on the power supply.

Safety note: Dry your hands thoroughly before handling the power supply.

Name ______________________ Date ______________________

9. Gel Electrophoresis

STUDENT ACTIVITY PAGE

8. Plug in the power supply, turn it on, and increase the voltage to about 100V.
9. Watch the electrodes carefully and record any color change you see near them. Be sure to match the correct color to the positive and negative electrodes.
10. Turn off your power supply.

EXTENSION OPTION

After completing the Procedure section, leave the water and indicator alone for a few minutes and then observe them again. What did you observe? Record your observations.

DATA COLLECTION AND ANALYSIS

Substance	Color before turning on power	Color after turning on power
Entire solution		
Solution surrounding cathode (negative electrode)		
Solution surrounding anode (positive electrode)		

CONCLUDING QUESTIONS

1. Bromothymol is an indicator that turns dark blue in bases and yellow in acids. What kind of substance is forming at the positive electrode?

____

____

____

____

2. What kind of substance is forming at the negative electrode?

____

____

____

____

Name ______________________ Date ______________________

9. Gel Electrophoresis

STUDENT ACTIVITY PAGE

3. When gel electrophoresis is performed on a large molecule and it breaks into a variety of charged particles, which electrode are the positively charged molecules attracted to?

____

____

____

____

4. When gel electrophoresis is performed on a large molecule and it breaks into a variety of charged particles, which electrode are the negatively charged molecules attracted to?

____

____

____

____

Follow-Up Activity

Conduct research to discover some of the kinds of molecules that are formed when DNA is broken into smaller pieces. Which of these molecules are positively charged and which are negatively charged? Present your findings to the class in a format approved by your teacher.

10. Building People

TEACHER RESOURCE PAGE

INSTRUCTIONAL OBJECTIVES

Students will be able to:

- discuss the various pros and cons of manipulating human traits with tools of biotechnology
- identify human traits that are genetically determined

NATIONAL SCIENCE EDUCATION STANDARDS CORRELATIONS

GRADES 5–8

Content standard	Bullet number	Content description	Bullet number(s)
A	1	Abilities necessary to do scientific inquiry	3, 4, 7
A	2	Understandings about scientific inquiry	1, 4
C	1	Structure and function in living systems	1, 2
C	2	Reproduction and heredity	3, 4
C	5	Diversity and adaptations of organisms	1
E	1	Abilities of technological design	1, 2, 4
E	2	Understandings about science and technology	4, 6

GRADES 9–12

Content standard	Bullet number	Content description	Bullet number(s)
A	1	Abilities necessary to do scientific inquiry	1–6
A	2	Understandings about scientific inquiry	1, 3, 6
C	1	The cell	1–4
C	4	Interdependence of organisms	3, 5
E	1	Abilities of technological design	1–5
E	2	Understandings about science and technology	1–3
F	6	Science and technology in local, national, and global challenges	4
G	3	Historical perspectives	3

VOCABULARY

- **ethical:** pertaining to behavior as decided by a group or society
- **moral:** relating to principles of right and wrong

10. Building People

TEACHER RESOURCE PAGE

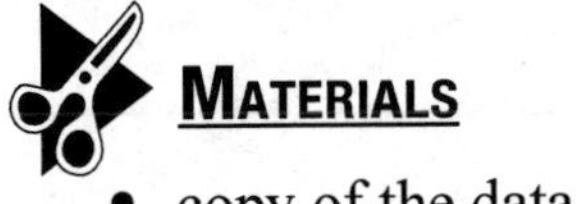

MATERIALS

- copy of the data table
- writing utensil

HELPFUL HINTS AND DISCUSSION

Time frame: 30 minutes
Structure: individuals
Location: classroom

The possibility exists that every human trait may be genetically determined. If this turns out to be true, it is possible that people may someday be able to do more than just select a hair color for their unborn child; they may be able to select a sense of humor for her or him as well. Other advances may make it possible for people to change some of their personal traits through some advanced form of gene therapy. Perhaps someday people will get up in the morning and take a pill that changes their eye color, or go to the doctor to get a shot that will change their gender. Today's students may have to make decisions about ideas like these in the future. This exercise will give them an opportunity to become familiar with some of the traits that we already know are genetically determined and to speculate about what they might do if the possibility of building a human or rebuilding themselves became available.

MEETING THE NEEDS OF DIVERSE LEARNERS

Students who need extra challenges should complete the Follow-Up Activity and the Extension Option. These students should be encouraged to think of more traits for their list. Students who need extra help will benefit from a discussion about the pros and cons of modifying their personal makeup.

SCORING RUBRIC

Students meet the standard for this activity by:

- ordering traits in all three columns of the data table
- answering the Concluding Questions
- contributing positively to a class discussion about genetic modification of humans
- adding descriptive terms to the data table

10. Building People

TEACHER RESOURCE PAGE

RECOMMENDED INTERNET SITES

- **Designer Babies: Eugenics Repackaged or Consumer Options?**
  www.accessmylibrary.com/coms2/summary_0286-30612482_ITM
- **Genetic Engineering**
  http://jp.senescence.info/thoughts/genetics.html
- **Human Reproduction and Genetic Ethics—The New Gene Technology and the Difference Between Getting Rid Of Illness and Altering People**
  www.geneticethics.org/back_issues/1_1_sutton.html

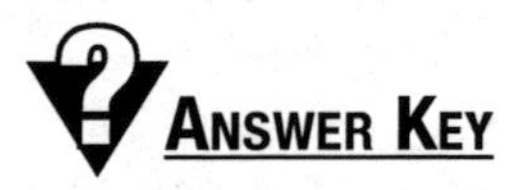

ANSWER KEY

1–6. Answers will vary widely depending on the way students ranked the traits on their lists. Encourage students to discuss their answers.

Name ______________________________ Date ______________________

10. Building People

STUDENT ACTIVITY PAGE

Objective

To explore the various traits that may be linked to the genetic makeup of people and the possible ethical and moral implications of modifying those traits with biotechnology

Before You Begin

Modern biotechnology is found in many different disciplines, but there is no question that the modification of the genetic makeup of organisms is one of its most powerful tools. As organisms, humans need to understand that we have a genetic makeup that can be modified as well. Whether this means fixing genetic problems, enhancing the human body, or creating people from scratch, there are a number of **ethical** and **moral** questions that should be answered. Is it okay to change the makeup of humans? Is it safe? Should parents be able to pick all the traits that will make up their unborn child? This exercise will allow you to think about the answers to some of those questions in detail.

Materials

- copy of the data table
- writing utensil

Procedure

Note: In the Data Collection and Analysis section, there is a table that has a list of twenty traits. These are all things that are possibly linked to your genetic makeup, and that you could possibly change someday using biotechnology.

1. In the column titled "You," number the traits in the order that you think are most important to how you are as a person overall. Use 1 as most important and 20 as least important.
2. In the column "Significant other," number the traits according to how desirable you would find them in a mate. For example, would you rather a potential mate have more patience than intelligence, or vice versa?
3. Assume that someday a genetic engineer could help you choose traits for your child. In the column "Your child," rank the traits in the order that you think would be most important for your child to have.
4. After ordering the traits in all three columns, fill in descriptive terms to indicate how you would like each trait to express itself. For example, beside "Height" you might indicate that you want to be 200 cm tall (about 6 feet 7 inches) if you feel too short or 150 cm (about 4 feet 11 inches) if you feel too tall. Beside "Sense of humor," if you feel like no one ever laughs at your jokes, you might choose to be "very funny." If people accuse you of being the class clown and you don't like it, you might wish to be "mostly serious."
5. Complete the Concluding Questions and then have a discussion with your classmates about the pros and cons of genetically modifying humans.

Name ______________________ Date ______________________

10. Building People

STUDENT ACTIVITY PAGE

EXTENSION OPTION

Identify which of the traits in the data table are known to be genetically determined.

DATA COLLECTION AND ANALYSIS

Trait	You	Significant other (girlfriend/boyfriend/spouse)	Your child
Build (musculature)			
Cheekbone shape			
Ear shape			
Eye color			
Eye shape			
Eyesight			
Hair color			
Health			
Hearing			
Height			
Intelligence			
Mouth shape			
Nose shape			
Overall attractiveness			
Patience			
Resistance to disease			
Sense of humor			
Skin color			
Skin complexion			
Weight			

Name ______________________ Date ______________________

10. Building People

STUDENT ACTIVITY PAGE

CONCLUDING QUESTIONS

Note: Many of these questions are opinion questions. Try to think carefully about your answers and think of reasons you would use to defend your answers.

1. What, if any, traits would you change about yourself? Why?

2. Look at the list of traits for your significant other. Would you want him or her to be exactly like you, slightly different, or completely different? Why?

3. Can you think of a real-world person who might be like the significant other you chose? If so, is it realistic to seek out a person like this? Explain.

4. What traits did you think were most important for your child? Are they similar to the traits you thought were important for yourself? Why or why not?

10. Building People

5. Is it right or wrong to modify yourself genetically? Why or why not?

____

____

____

____

6. Is it right or wrong to modify your child or children? Why or why not?

____

____

____

____

FOLLOW-UP ACTIVITY

Design a pet that has traits from a wide variety of animals. List the animals you chose traits from and the traits you chose. Would you buy such a pet if you could? Present your findings to the class in a format approved by your teacher.

11. Running a Gel

TEACHER RESOURCE PAGE

INSTRUCTIONAL OBJECTIVES

Students will be able to:

- perform gel electrophoresis on a sample
- compare results from gel electrophoresis
- identify an unknown using gel electrophoresis

NATIONAL SCIENCE EDUCATION STANDARDS CORRELATIONS

GRADES 5–8

Content standard	Bullet number	Content description	Bullet number(s)
A	1	Abilities necessary to do scientific inquiry	3, 4, 7
A	2	Understandings about scientific inquiry	1, 4
E	2	Understandings about science and technology	4, 6
F	5	Science and technology in society	3, 7

GRADES 9–12

Content standard	Bullet number	Content description	Bullet number(s)
A	1	Abilities necessary to do scientific inquiry	1–6
A	2	Understandings about scientific inquiry	1, 3, 6
B	3	Chemical reactions	1
E	1	Abilities of technological design	1–5
E	2	Understandings about science and technology	1–3
G	3	Historical perspectives	3

VOCABULARY

- **gel electrophoresis:** a process for separating large organic molecules into smaller component molecules using an electric field directed through a gel

11. Running a Gel

TEACHER RESOURCE PAGE

MATERIALS

- electrophoresis unit
- power supplies
- dyes:
  - bromophenol blue
  - methyl green
  - methyl orange
  - orange G
  - pyronin Y
  - safranin
  - xylene cyanol
  - unknown dye (a mixture of any two or three of the dyes above)
- glycerol
- 1.0 % agarose in 1× TAE solution
- ethidium bromide
- 1× TAE solution
- micropipettes
- ruler
- stirring rod
- graduated cylinder
- test tubes
- goggles
- lab aprons
- gloves
- distilled water

= Safety icon

HELPFUL HINTS AND DISCUSSION

Time frame: 90 minutes
Structure: individuals or partners
Location: classroom

To prepare the dyes for the lab, use the following procedure. Repeat for each dye.

1. Weigh out 0.2 g of dye and dissolve (as much as possible) in 10 ml of distilled water.
2. Fill a graduated cylinder with 18 ml of distilled water.
3. Add 1 ml of solution from step 1 to the graduated cylinder.
4. Add 1 ml of glycerol to the graduated cylinder and stir well with a stirring rod. Store in labeled test tubes. Excess dye solution can be refrigerated for later use.

The unknown dye in this lab can be a combination of any of the other dyes. If students are running a gel for the first time, it would be preferable if the unknown was a combination of only two of the dyes. You may not have access to all of the dyes listed. If this is the case, use as many as you can obtain and seek suitable substitutions. To decrease the time of the lab, it would also be preferable if you prepared the agarose and TAE if students have not done either before.

MEETING THE NEEDS OF DIVERSE LEARNERS

Students who need extra challenges should complete the Follow-Up Activity and the Extension Option. Students who need extra help will benefit from a pre-lab discussion showing how to compare dye bands for the purpose of matching a sample to an unknown. They should also be allowed to work with a partner. You may wish to create a diagram of the electrophoresis unit for students unable to draw one themselves, and have them label the diagram.

11. Running a Gel

SCORING RUBRIC

Students meet the standard for this activity by:

- drawing accurately scaled and labeled diagrams
- identifying the components of the unknown dye
- following safety requirements
- collecting accurate data and presenting it in a clear manner
- answering the Concluding Questions

RECOMMENDED INTERNET SITES

- **Colorado State University—Biotechnology and Genetic Engineering: Agarose Gel Electrophoresis of DNA**
  www.vivo.colostate.edu/hbooks/genetics/biotech/gels/agardna.html
- **Molecular Biology Cyberlab—Experiment 2: Gel Electrophoresis of DNA**
  www.life.illinois.edu/molbio/geldigest/electro.html
  (Scroll down to the section titled "What Is a Gel?" where you will find links to pictures of the process of preparing a gel.)
- **The University of Tennessee at Knoxville—Agarose Gels**
  www.bio.utk.edu/mycology/Techniques/mt-agarose_gels.htm
  (Scroll down to the section titled "2. Gel Preparation," where you will find detailed directions for preparing agarose gel.)
- **The University of Tennessee at Knoxville—Gel Electrophoresis**
  http://web.utk.edu/~khughes/GEL/sld001.htm

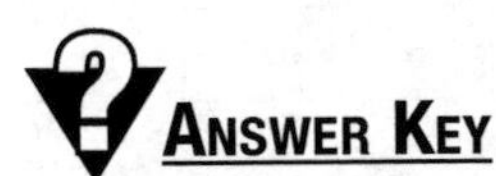

ANSWER KEY

1. Those particles have a positive electrical charge.
2. Those particles have a negative electrical charge.
3. Answers will vary based on the makeup of the unknown dye. Students should be able to match the dye bands by their locations.
4. The dyes are made of a number of different molecules that have different weights and travel through the gel at different rates.

Name ______________________ Date ______________________

11. Running a Gel

STUDENT ACTIVITY PAGE

OBJECTIVE

To perform gel electrophoresis on a series of known samples for comparison to an unknown sample for the purpose of identification

BEFORE YOU BEGIN

Gel electrophoresis is a process that allows molecules such as DNA to be separated into smaller molecules. The final product of electrophoresis allows the user to find many properties related to the molecule that is being tested. Gel electrophoresis can be used to make a profile for a biological compound of unknown origin. Then that profile can be compared to a database of known profiles. Such tests are used to identify genetic diseases, to identify a criminal, to identify substances, or to establish paternity. Gel electrophoresis is one of the core tools of biotechnology and has many variants. The gel that the samples are run though can be made of many different materials. The number of dyes and substances that could be put into the gel for processing is almost unlimited.

MATERIALS

- electrophoresis unit
- dyes:
  - bromophenol blue
  - methyl green
  - methyl orange
  - orange G
  - pyronin Y
  - safranin
  - xylene cyanol
  - unknown dye (a mixture of any two or three of the dyes above)
- power supplies
- 1.0 % agarose in 1× TAE solution
- ethidium bromide
- 1× TAE solution
- micropipettes
- ruler
- goggles
- lab apron
- gloves

PROCEDURE

Safety note: Heated agarose mixture can have superheated spots that can burn you. Make sure you're wearing gloves, and handle the agarose carefully.

1. Set up the electrophoresis unit and power supply as directed.
2. Place a single comb in the center of the gel tray and seal the tray ends.
3. Fill the gel tray with hot agarose mixture. Let stand for at least 10 minutes. Do not touch or disturb the gel.

Name ______________________ Date ______________________

11. Running a Gel

STUDENT ACTIVITY PAGE

4. While you're waiting for the gel to harden, draw a diagram of your electrophoresis unit and label the wells with numbers (1 through 8).
5. After the gel has hardened, unseal the tray ends and cover the gel with TAE solution.
6. Remove the comb.
7. Fill each well with 10 to 15 ml of a different dye sample using the micropipettes. Add only one dye to each well. Write the name of the dye you used for each well next to the same number in the data table in the Data Collection and Analysis section and on your diagram.
8. Turn on your power supply to approximately 100 V. Let the gel run for at least 10 minutes.
9. Turn off the power supply and unplug it.
10. Using a ruler, measure the distance from the end of the gel tray to each dye band. Record the distance in your data table.
11. In your diagram, add a drawing for each dye that is to scale. Show the locations of the positive and negative terminals in each of your drawings. This will help you identify the direction of motion of the variously charged particles in the tray.

Extension Option

Find some other biological dyes and run the lab again. Try to make your measurements and drawings as accurate as possible.

Data Collection and Analysis

Well	Dye name	Distance to each dye band from end of tray
1		
2		
3		
4		
5		
6		
7		
8		

Name ______________________ Date ______________________

11. Running a Gel

STUDENT ACTIVITY PAGE

Concluding Questions

1. Some of the dye moved toward the negative terminal. What does this indicate?

2. Some of the dye moved toward the positive terminal. What does this indicate?

3. What dye or dyes made up the unknown? How could you tell?

4. Why do the dyes separate into a number of bands?

Follow-Up Activity

Research to find out what the biological dyes used in this lab are used for in other disciplines of biotechnology. Present your findings to the class in a format approved by your teacher.

12. To Clone, or Not to Clone?

TEACHER RESOURCE PAGE

INSTRUCTIONAL OBJECTIVES

Students will be able to:

- describe the various viewpoints held about a variety of differing cloning uses
- describe some of the modern uses of cloning

NATIONAL SCIENCE EDUCATION STANDARDS CORRELATIONS

GRADES 5–8

Content standard	Bullet number	Content description	Bullet number(s)
A	1	Abilities necessary to do scientific inquiry	3, 4, 7
A	2	Understandings about scientific inquiry	1, 4
C	1	Structure and function in living systems	1, 2
C	2	Reproduction and heredity	3, 4
C	3	Regulation and behavior	1, 4
C	5	Diversity and adaptations of organisms	1
E	1	Abilities of technological design	1, 2, 4
E	2	Understandings about science and technology	4, 6
F	5	Science and technology in society	3, 7

GRADES 9–12

Content standard	Bullet number	Content description	Bullet number(s)
A	1	Abilities necessary to do scientific inquiry	1–6
A	2	Understandings about scientific inquiry	1, 3, 6
C	1	The cell	1–4
C	2	Molecular basis of heredity	1, 3
C	4	Interdependence of organisms	3, 5
E	1	Abilities of technological design	1–5
E	2	Understandings about science and technology	1–3
F	6	Science and technology in local, national, and global challenges	4
G	3	Historical perspectives	3

VOCABULARY

- **cloning:** the process of reproducing organisms so that they have a genetic makeup that is identical to that of the original organism
- **vegetative propagation:** the process of cloning plants from cuttings

12. To Clone, or Not to Clone?

TEACHER RESOURCE PAGE

MATERIALS

- Internet access
- notebook

HELPFUL HINTS AND DISCUSSION

Time frame: one week for research
Structure: individuals
Location: classroom or home

This activity will allow students to explore the various ethical viewpoints surrounding the ways that cloning is currently used and the ways that cloning might eventually be used. Students will find five modern uses of cloning and then investigate the various viewpoints that individuals or groups have about those uses. Special care should be taken to remind students that this is an opportunity to explore the opinions and beliefs of others. They are not required to give their own viewpoint and should focus on collecting information to summarize and share instead.

Since students will compile a bibliography of their sources, consult a language arts teacher as to which format is generally used in your school, and have bibliographic reference materials handy so that students can look at examples of how to list each type of source they use.

MEETING THE NEEDS OF DIVERSE LEARNERS

Students who need extra challenges should complete the Follow-Up Activity and the Extension Option. These students should also be encouraged to make a set of guidelines to keep their observations free of bias. Students who need extra help may also benefit from some assistance in choosing their five cloning topics. They should also be allowed to work with a partner if necessary.

SCORING RUBRIC

Students meet the standard for this activity by:

- thoroughly addressing five cloning topics
- recording and summarizing major opinions about their topics
- presenting a short presentation about one topic they found interesting
- answering the Concluding Questions

RECOMMENDED INTERNET SITES

- **HowStuffWorks—How Cloning Works**
  http://science.howstuffworks.com/genetic-science/cloning.htm
- **Society, Religion and Technology Project: Therapeutic Uses of Cloning and Embryonic Stem Cells**
  www.srtp.org.uk/clonin50.htm

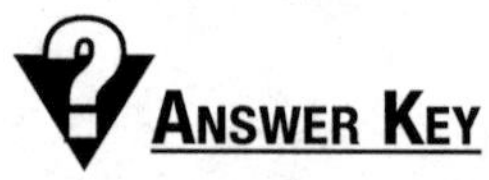

ANSWER KEY

1–4. Answers will vary based on the cloning topics students chose.

Name ____________________ Date ____________________

12. To Clone, or Not to Clone?

Student Activity Page

Objective

To become familiar with the various viewpoints held by people with respect to the many uses and possible uses of cloning technology

Before You Begin

For a long time, people have used a process called **vegetative propagation** to clone plants. The process generally requires that a piece of plant is cut away and then either replanted or encouraged to grow new roots. This new plant is a clone of the original because it has an identical genetic makeup to the original plant. There are forms of modern **cloning,** however, that rely on the direct manipulation of genetic material to produce the clones. This allows scientists to produce clones of many different organisms. Scientists may also modify the genetic material of the organisms in the process to produce new organisms or only parts of organisms. Biotechnology has provided the tools to allow cloning to take place. The question that many people want answered is: Is it right to clone?

Materials

- Internet access or other resources
- notebook

Procedure

1. Using the Internet or print sources such as newspapers, magazines, books, journals, and so forth, determine five modern uses of cloning. You may also choose to include a predicted use of cloning that is not currently in use. List your sources and the uses of cloning in the Data Collection and Analysis section.
2. Almost any aspect of cloning will have people who support its use and people who are against its use. Research and summarize the various pro and con opinions that people hold about the five cloning uses you chose. You will find that religious organizations, philosophers, and medical professionals are all groups that have widely available opinions about cloning.
3. As you go, record each of your sources in the Data Collection and Analysis section to turn in to your teacher. Follow the bibliography format preferred at your school.
4. Write a paper that summarizes your findings. Split the paper into five distinct sections, one for each cloning use. Be sure to include a list of your sources.
5. Using the information for the cloning topic that you found most interesting, prepare a brief presentation to share the pros and cons with your classmates. Limit yourself to one topic only.

Extension Option

Read about the cloning of plants from cuttings. Are there any groups that are particularly opposed to this sort of cloning?

Name ______________________ Date ______________________

12. To Clone, or Not to Clone?

STUDENT ACTIVITY PAGE

DATA COLLECTION AND ANALYSIS

Use 1 (describe): ______________________

____

Source of information: ______________________

____

Use 2 (describe): ______________________

____

Source of information: ______________________

____

Use 3 (describe): ______________________

____

Source of information: ______________________

____

Use 4 (describe): ______________________

____

Source of information: ______________________

____

Use 5 (describe): ______________________

____

Source of information: ______________________

____

Name ______________________________ Date ______________________

12. To Clone, or Not to Clone?

STUDENT ACTIVITY PAGE

CONCLUDING QUESTIONS

1. What were some common objections to the use of cloning?

____

____

____

____

2. What were some common arguments in support of cloning?

____

____

____

____

3. Choose one of the arguments for cloning. What logical reasons support this argument?

____

____

____

____

4. Choose one of the arguments against cloning. What logical reasons support this argument?

____

____

____

____

FOLLOW-UP ACTIVITY

Look for some examples of cloning technology being used in works of fiction, whether in literature, movies, or on television. How is cloning treated by the people who created the fiction? Are they generally for it or against it? Present your findings to the class in a format approved by your teacher.

13. Was It Alive?

Teacher Resource Page

Instructional Objectives

Students will be able to:

- identify the presence of protein in a substance
- understand the uses of proteins as tools of biotechnology

National Science Education Standards Correlations

Grades 5–8

Content standard	Bullet number	Content description	Bullet number(s)
A	1	Abilities necessary to do scientific inquiry	3, 4, 7
A	2	Understandings about scientific inquiry	1, 4
C	1	Structure and function in living systems	1, 2
C	2	Reproduction and heredity	3, 4
F	5	Science and technology in society	3, 7

Grades 9–12

Content standard	Bullet number	Content description	Bullet number(s)
A	1	Abilities necessary to do scientific inquiry	1–6
A	2	Understandings about scientific inquiry	1, 3, 6
B	3	Chemical reactions	1, 5
C	1	The cell	1–4
C	2	Molecular basis of heredity	1, 3
C	4	Interdependence of organisms	3, 5
E	1	Abilities of technological design	1–5
E	2	Understandings about science and technology	1–3
G	3	Historical perspectives	3

Vocabulary

- **Biuret solution:** a solution used to identify the likely presence of proteins in a sample
- **proteins:** organic compounds made of amino acids that are found in all living organisms

13. Was It Alive?

TEACHER RESOURCE PAGE

MATERIALS

- ten test tubes
- test-tube rack
- Biuret solution
- marking pencil or masking tape for labeling tubes
- 10-ml graduated cylinder
- apple
- blueberries
- flower petals
- pepper
- salt
- raw steak
- strawberries
- sugar
- vegetable oil
- distilled water
- goggles
- lab aprons
- gloves

HELPFUL HINTS AND DISCUSSION

Time frame: 60 minutes
Structure: individuals or partners
Location: classroom

This activity is relatively simple, but there are some safety issues. **The Biuret solution does contain sodium hydroxide, so students should wear goggles and gloves and should not allow the solution to get on their skin. If contact is made, they should wash the affected area under running water immediately.** The Biuret solution reacts with the amino groups (NH_2) in the proteins that do not appear in carbohydrates, fats, and nucleic acids. Although this is not the definitive test for proteins, it is a very simple tool that biotechnologists can use as a starting point when studying new substances.

MEETING THE NEEDS OF DIVERSE LEARNERS

Students who need extra challenges should complete the Follow-Up Activity and the Extension Option. They should also be encouraged to supervise the safety of the work environment. Students who need extra help will benefit from a clarification of the fact that protein isn't only found in muscle tissue, as some people believe.

SCORING RUBRIC

Students meet the standard for this activity by:

- answering the Concluding Questions
- correctly determining which substances contain proteins
- accurately recording observations
- carefully handling lab equipment and chemicals

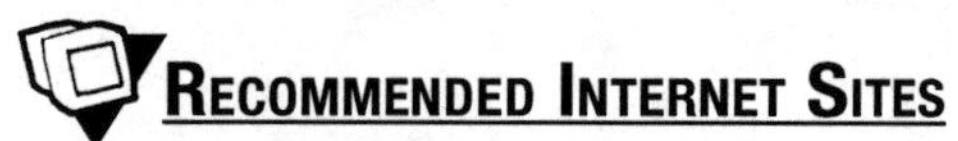

13. Was It Alive?

TEACHER RESOURCE PAGE

Recommended Internet Sites

- **Human Protein Atlas**
  www.proteinatlas.org
- **Huntington's Outreach Project for Education, at Stanford—An Introduction to Proteins**
  http://hopes.stanford.edu/basics/proteins/p0.html
- **Protein Data Bank—A Resource for Studying Biological Macromolecules**
  www.rcsb.org/pdb/home/home.do

Answer Key

1. The apples, blueberries, flower petals, pepper, steak, and strawberries should all test positive for proteins.
2. The salt, sugar, vegetable oil, and distilled water should not test positive for proteins.
3. Answers will vary.
4. Sugar does not contain an amino group.

Name ______________________________ Date ______________________________

13. Was It Alive?

STUDENT ACTIVITY PAGE

OBJECTIVE

To determine whether or not a particular material contains proteins

BEFORE YOU BEGIN

Proteins are found in all living things and do many jobs. They are a core substance used in many of the fields of biotechnology. In the human body alone, proteins are essential to the health of our muscles, hair, skin, and fingernails. They perform a number of tasks as enzymes, immune-function molecules, and hormones. They are even necessary for the contraction of the muscles that let us move and even breathe. Testing a material for proteins is a pretty good way to see if it is alive or if it was ever alive. In this activity, you will add **Biuret solution** to different samples. If the solution turns from blue to purple, there is a high likelihood that the sample contains protein.

MATERIALS

- ten test tubes
- test-tube rack
- Biuret solution
- marking pencil or masking tape for labeling tubes
- 10-ml graduated cylinder
- apple
- blueberries
- flower petals
- pepper
- salt
- raw steak
- strawberries
- sugar
- vegetable oil
- distilled water
- goggles
- lab apron
- gloves

PROCEDURE

1. Number your test tubes 1 through 10.
2. Place 1 or 2 ml of distilled water in tube number 1. The water will act as your control.
3. Place a small amount, about the size of a pencil eraser, of each sample into its own test tube. Use 1 or 2 ml for the liquid samples. In the data table in the Data Collection and Analysis section, write the name of each sample next to the number of the tube you placed it in.
4. In your data table, record your guess as to whether or not the sample contains protein.
5. Pour 4 to 5 ml of Biuret solution into each test tube.
6. Observe the test tubes and watch for any color change in the Biuret solution. Record any color you see in your data table. If the solution turns from blue to purple, the sample probably contains protein.

Name ______________________________ Date ______________________

13. Was It Alive?

STUDENT ACTIVITY PAGE

Extension Option

Under the guidance of your teacher, remove some dead skin fragments, fingernail clippings, and so forth from your body and test for the presence of proteins.

Data Collection and Analysis

Tube	Substance	Do you think it will test positive for proteins? (Y/N)	Color of substance and Biuret solution	Positive test for proteins? (Y/N)
1	Distilled water			
2				
3				
4				
5				
6				
7				
8				
9				
10				

Concluding Questions

1. Which of the substances you tested contained proteins?

____

____

____

____

Name ______________________________ Date ______________________

13. Was It Alive?

STUDENT ACTIVITY PAGE

2. Which of the substances you tested did not contain proteins?

____

____

____

____

3. Were any of the substances that contained proteins a surprise to you? Why or why not?

____

____

____

____

4. The formula for sugar is $C_{12}H_{22}O_{11}$. Why is sugar unlikely to test positive for the presence of proteins?

____

____

____

____

FOLLOW-UP ACTIVITY

Research to find out how many different proteins there are in the human body and where they are found. Present your findings to the class in a format approved by your teacher.

14. The Community and Genetically Modified Foods

TEACHER RESOURCE PAGE

INSTRUCTIONAL OBJECTIVES

Students will be able to:

- explain the various viewpoints of community members with respect to genetically modified foods
- determine the various stakeholders in the use of genetically modified foods
- understand the various opinions of community members with respect to the use of genetically modified foods

NATIONAL SCIENCE EDUCATION STANDARDS CORRELATIONS

GRADES 5–8

Content standard	Bullet number	Content description	Bullet number(s)
A	1	Abilities necessary to do scientific inquiry	3, 4, 7
A	2	Understandings about scientific inquiry	1, 4
C	1	Structure and function in living systems	1, 2
C	2	Reproduction and heredity	3, 4
C	3	Regulation and behavior	1, 4
C	5	Diversity and adaptations of organisms	1
E	1	Abilities of technological design	1, 2, 4
E	2	Understandings about science and technology	4, 6
F	5	Science and technology in society	3, 7

GRADES 9–12

Content standard	Bullet number	Content description	Bullet number(s)
A	1	Abilities necessary to do scientific inquiry	1–6
A	2	Understandings about scientific inquiry	1, 3, 6
C	1	The cell	1–4
C	2	Molecular basis of heredity	1, 3
C	4	Interdependence of organisms	3, 5
E	1	Abilities of technological design	1–5
E	2	Understandings about science and technology	1–3
F	6	Science and technology in local, national, and global challenges	4
G	3	Historical perspectives	3

14. The Community and Genetically Modified Foods

TEACHER RESOURCE PAGE

VOCABULARY

- **biotechnology:** the use or modification of organisms for human purposes
- **genetic engineering:** a process of manipulating an organism's genes
- **genetically modified organism:** an organism that has had some part of its genetic makeup altered by genetic engineering

MATERIALS

- Internet access
- notebook
- presentation space
- Optional: wardrobe pieces appropriate to the roles; for example, a cook's hat, plastic pitchfork or shovel, reporter's notebook, etc.

HELPFUL HINTS AND DISCUSSION

Time frame: one week for research and preparation, plus class time for presentations
Structure: individuals or groups, and class
Location: classroom, library

This activity allows students to act out the roles of community members in an area where genetically modified foods are being grown. They will take on the roles of individuals and will then try to work for an outcome that best benefits the group represented by that role. For example, if the student takes on the role of farmer, then he or she will try to advocate for an outcome that best benefits farmers in the community. Perhaps the student will point out the positive benefits of higher crop yields, pest and drought resistance, increased growing season length, and so forth.

The final "town meeting" will be a place for students to present their opinions and facts in support of those opinions. A committee of three or five students (preferably an odd number to prevent a tie vote) will vote when the presentation is over as to whether or not to allow the genetically modified food crop to be grown in the area. A good extension option is to repeat the activity and to have the students change roles so committee members can take on roles as community members. Some suggested roles:

- biologist from local university
- cook concerned with using local produce
- Department of Agriculture employee
- environmentalist who is opposed to genetically modified crops
- farmer who will be growing the crop
- farming neighbor near farm that will grow the new crop
- seed supplier of the new crop
- parent concerned with safety of feeding genetically modified foods to his or her children
- police chief concerned with potential violence directed at farmer
- reporter trying to cover all angles of the story
- student assigned to find out about genetically modified foods for a class
- supermarket owner concerned with stocking genetically modified foods

(continued on next page)

14. The Community and Genetically Modified Foods

TEACHER RESOURCE PAGE

You may wish to think of other community members who may be affected by the introduction of crops to the local scene. If you would like to limit the number of individuals who have to speak at the town meeting, you may form groups or partners for each community group. The length of the final community meeting will be limited by your schedule and the number of students you wish to speak. A rubric chart is provided in the Data Collection and Analysis section to help students evaluate each presentation. Depending on the number of individuals/groups who present, you may need to make extra copies. You may direct each student in the audience to fill out a rubric, or allow groups to fill out one chart together. You may also use this chart to help you grade participation for this project.

MEETING THE NEEDS OF DIVERSE LEARNERS

Students who need extra challenges should complete the Follow-Up Activity and the Extension Option. These students might also make good members of the voting committee the first time you use this activity, since they will have to keep track of a large amount of data. Students who need extra help should be allowed to work with a partner to do their research if necessary. It may be helpful to invite one group or individual to present first as a "test run," then guide students through filling out the rubric for that presentation.

SCORING RUBRIC

Students meet the standard for this activity by:

- collecting and presenting facts or opinions that are relevant to the group they are representing
- answering the Concluding Questions
- politely listening to opposing viewpoints while remaining civil
- remaining impartial (if on voting committee)

RECOMMENDED INTERNET SITES

- **Monsanto Company home page**
  www.monsanto.com
- **ProQuest—Genetically Modified Foods: Harmful or Helpful?**
  www.csa.com/discoveryguides/gmfood/overview.php
- **World Health Organization—20 Questions on Genetically Modified (GM) Foods**
  www.who.int/foodsafety/publications/biotech/20questions/en/

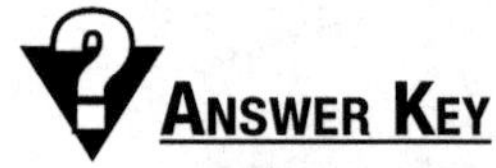

ANSWER KEY

1. Farmers, businesses, and consumers all stand to gain from the introduction of genetically modified crops. Students will provide a wide range of other reasons.
2. The environmentalist who is opposed to genetically modified crops will not be happy with the introduction of the GM crop. Some farmers who choose non-modified crops may also be unable to match the production of modified crops. Students will provide a wide range of other reasons.

3–5. Answers will vary.

Name ______________________ Date ______________________

14. The Community and Genetically Modified Foods

STUDENT ACTIVITY PAGE

OBJECTIVE

To research and understand the wide variety of viewpoints held by community members with respect to the growing of genetically modified foods

BEFORE YOU BEGIN

Biotechnology may sound like something that is only used by scientists. After a product is produced using biotechnology, however, it has the ability to affect all of the people who come in contact with it. If, for example, every farmer decided to only raise tomatoes, the effect on the rest of the community would become clear in a short amount of time. But what if one farmer introduced a **genetically modified** crop to his or her farm in a small community? What concerns would other farmers have? What responsibility does the local government have? This activity will help you answer those questions and many more.

MATERIALS

- Internet access
- notebook
- costumes (optional)

PROCEDURE

1. As a class, make a list of all of the kinds of people found in a community (such as doctors, reporters, police officers, teachers, students, parents, and so forth.)
2. After completing the list, determine which of the community members would be most affected by the introduction of a genetically modified crop into the local area. Your teacher will help you determine how many groups will be represented.
3. Your teacher will assign you a role. Then you will research the various concerns or wishes that may be held by the group you represent. Some students will be on a town council that has the authority to vote on the issues of introducing genetically modified crops to the local area. These students will research genetically modified crops.
4. Hold a town meeting that will be moderated by the town council or your teacher. Each group will be given the same amount of time to present their opinions and facts. Each group should have about 5 minutes to make their point, but may be given more time by the moderator. All groups will stay after their presentation to answer any questions the other community members may have.
5. Use the rubric in the Data Collection and Analysis section to rate the presentations you observe. For each category, give each individual or group a rating from 1 to 5, with 5 being best and 1 being worst. Refer to your rankings when it is time to vote. You will not give yourself a ranking; however, it is important to keep in mind the requirements to achieve higher rankings in each category as you prepare your presentation.
6. After all individuals/groups have presented their information, the town council will vote to decide whether or not to allow the genetically modified crops to be grown in the community.

Name ______________________ Date ______________________

14. The Community and Genetically Modified Foods

STUDENT ACTIVITY PAGE

EXTENSION OPTION

Change roles and repeat the activity. Be sure that the students who served on the town council have an opportunity to play a different community member role.

DATA COLLECTION AND ANALYSIS

Look over the following ranking system before evaluating each group. Write the name of each person/group in the diagonal line on the chart. Then, after each group presents their argument, rank the presentation from 1 to 5, with 5 being best.

- **5:** Excellent, thorough work that fully met the requirements of the category and included aspects that exceeded requirements.
- **4:** Fully met the requirements of the category.
- **3:** Met most of the requirements of the category. Research/effort could have more complete.
- **2:** Met some of the requirements of the category. Needed more research/effort.
- **1:** Did not meet requirements. This presenter/group may have not presented many facts, relied too heavily on strong opinions, or presented their own opinions rather than opinions that might be held by the community member(s) they were supposed to portray.

Group presented opinions effectively so that they were well understood.						
Facts supported the group's opinion and seemed well researched.						
Facts and opinions made sense given the group's assigned role.						
Group portrayed assigned role in a believable manner.						
Group's answers to audience questions were thorough and referenced research.						
Overall, the presentation was interesting, informative, and persuasive.						

Name ______________________________ Date ______________________________

14. The Community and Genetically Modified Foods

Student Activity Page

Concluding Questions

1. Which group or groups seem to have the most to gain from the introduction of genetically modified crops to a local farm? Why?

____

____

____

____

2. Which group or groups seem to have the most to lose from the introduction of genetically modified crops to a local farm? Why?

____

____

____

____

3. What were five benefits cited by groups interested in bringing genetically modified crops to this community?

____

____

____

____

4. What were five drawbacks cited by groups opposed to bringing genetically modified crops to this community?

____

____

____

____

14. The Community and Genetically Modified Foods

5. If your committee voted to bring in the genetically modified crops, what reasons were given for this decision? What reasons were given if the decision was to not bring in genetically modified crops?

____

____

____

____

FOLLOW-UP ACTIVITY

Find some accounts from a news source of events like these taking place in real life. What groups are mentioned in the story? Present your findings to the class in a format approved by your teacher.

15. Oil Cleanup with Bacteria

TEACHER RESOURCE PAGE

Instructional Objectives

Students will be able to:

- understand the concept of bioremediation
- understand bioremediation as a tool of biotechnology
- evaluate the efficiency of bioremediation under differing conditions

National Science Education Standards Correlations

Grades 5–8

Content standard	Bullet number	Content description	Bullet number(s)
A	1	Abilities necessary to do scientific inquiry	3, 4, 7
A	2	Understandings about scientific inquiry	1, 4
C	1	Structure and function in living systems	1, 2
C	2	Reproduction and heredity	3, 4
C	3	Regulation and behavior	1, 4
E	2	Understandings about science and technology	4, 6
F	5	Science and technology in society	3, 7

Grades 9–12

Content standard	Bullet number	Content description	Bullet number(s)
A	1	Abilities necessary to do scientific inquiry	1–6
A	2	Understandings about scientific inquiry	1, 3, 6
B	3	Chemical reactions	1, 5
C	1	The cell	1–4
C	2	Molecular basis of heredity	1, 3
C	4	Interdependence of organisms	3, 5
E	1	Abilities of technological design	1–5
E	2	Understandings about science and technology	1–3
F	6	Science and technology in local, national, and global challenges	4
G	3	Historical perspectives	3

Vocabulary

- **bioremediation:** the process of breaking down or removing toxins from the environment using living organisms or components of living organisms, such as plants, enzymes, fungi, and any of the many kinds of microorganisms

15. Oil Cleanup with Bacteria

Teacher Resource Page

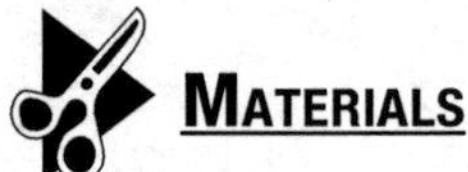

Materials

- chlorine-free bottled water (gallon jugs)
- bucket
- eyedroppers
- incubator
- motor oil
- petri dishes—eight per group
- plant fertilizer
- salt
- scales
- soil
- sugar
- weighing paper
- goggles for each student
- lab aprons
- gloves
- marking pencil or tape for labeling petri dishes

Helpful Hints and Discussion

Time frame: three to seven days
Structure: partners
Location: classroom

This activity will allow students to see the effectiveness of using bacteria from soil to break down oil. This is similar to how bacteria are used to break down the oil in oil spills, although the bacteria used by professionals are of a few specific kinds that have been grown for that purpose. You will provide the bacteria by collecting some soil, preferably from a location where it is clear the soil is fertile. You will need to add a shovelful of soil to a bucket and then to rinse the soil with a gallon of spring water. Do not use tap water, because the chlorine may decrease the amount of living bacteria. Mix the water and soil around and then leave it standing overnight. Decant the liquid back into the gallon jug and try not to get any of the solids back into the jug. This is your supply of bacteria-filled water. You may need to repeat this process if you have a large number of students. You may feel it necessary to filter the water before giving it to students to use if there are too many solids. They will then add oil to the bacteria-containing water and watch how the oil behaves under a number of different arrangements.

Meeting the Needs of Diverse Learners

Students who need extra challenges should complete the Follow-Up Activity and the Extension Option. These students should also be encouraged to think of alternative variables to manipulate in their research of the oil-eating bacteria. Students who need extra help may benefit from a step-by-step review of the procedure.

15. Oil Cleanup with Bacteria

TEACHER RESOURCE PAGE

SCORING RUBRIC

Students meet the standard for this activity by:

- correctly using lab equipment
- making clear and accurate observations and recording them appropriately
- drawing conclusions about the effects of manipulation variables in an experiment
- answering the Concluding Questions

RECOMMENDED INTERNET SITES

- **Transgalactic Ltd—What Is Bioremediation?**
  www.bionewsonline.com/w/what_is_bioremediation.htm
- **United States Environmental Protection Agency—A Citizen's Guide to Bioremediation**
  www.epa.gov/swertio1/download/citizens/bioremediation.pdf
- **United States Environmental Protection Agency—Bioremediation of Exxon Valdez Oil Spill**
  www.epa.gov/history/topics/valdez/01.htm
- **U.S. Geological Survey—Bioremediation: Nature's Way to a Cleaner Environment**
  http://water.usgs.gov/wid/html/bioremed.html

ANSWER KEY

1. Answers will vary depending on the bacteria found in the soil sample. However, the dish in the incubator with the fertilizer should show a large change.
2. Answers will vary depending on the bacteria found in the soil sample. However, the dishes with just oil and water and the dishes with oil, water, and salt will show less change than the others.
3. In general, the samples in the incubator should be broken down more.
4. Some warmth and fertilizer should help the process of breaking down the oil.

Name ______________________________ Date ______________________________

15. Oil Cleanup with Bacteria

STUDENT ACTIVITY PAGE

OBJECTIVE

To determine the effectiveness of bacteria as oil-eaters under several different conditions

BEFORE YOU BEGIN

Bioremediation is a tool of biotechnology that breaks down or removes toxins from the environment. This is done using living organisms or components of living organisms such as plants, enzymes, fungi, and any of the many kinds of microorganisms. Bioremediation has been used in a variety of ways. One of the first, large-scale uses was to help clean up an oil spill in Prince William Sound, Alaska. The approach was to encourage the growth of naturally occurring bacteria by adding fertilizers. The fertilizers provided nutrients in the form of phosphorus and nitrogen. These were necessary to allow the bacteria to break down the oil in their digestive process. The long-chain hydrocarbon molecules are broken down into simpler, less toxic by-products, such as carbon dioxide and water. In this activity, you will experiment with the conditions that might encourage oil-eating bacteria to complete the breakdown process even faster.

MATERIALS

- bacteria-contaminated water
- eyedroppers
- incubator
- motor oil
- eight petri dishes
- plant fertilizer
- salt
- scales
- sugar
- weighing paper
- goggles
- lab apron
- gloves
- marking pencil or tape for labeling petri dishes

PROCEDURE

1. Get eight petri dishes. Label four of them "Incubator 1" through "Incubator 4." Label the other four as "Shelf 1" through "Shelf 4." Put your initials and your partner's initials on each dish. As you add materials, be sure to record the contents of each dish in the appropriate data table in the Data Collection and Analysis section.
2. Fill each of the dishes half full of the "bacteria water" supplied by your teacher.
3. Add 15 drops of motor oil to each dish. *Note:* Do not add anything else to the dishes marked Incubator 1 and Shelf 1. Set those dishes aside and record their contents in the data tables.
4. Add 1 gram of fertilizer to the dishes marked Incubator 2 and Shelf 2. Record their contents in the data tables.
5. Add 1 gram of sugar to the dishes marked Incubator 3 and Shelf 3. Record their contents.

Name ____________________ Date ____________________

15. Oil Cleanup with Bacteria

STUDENT ACTIVITY PAGE

6. Add 1 gram of salt to the dishes marked Incubator 4 and Shelf 4. Record their contents.
7. Inspect each dish. Record your observations for each beside "Day 1" in the appropriate data table.
8. Store the "Shelf" dishes somewhere at room temperature and put the "Incubator" dishes in the incubator. Set the incubator to 37°C.
9. Check each dish every day for seven days. Record your observations about the state of the oil in each dish in the appropriate data table. If your experiment goes over a weekend, renumber your data tables accordingly and take readings for seven total days.

EXTENSION OPTION

Repeat this experiment and modify some of the variables. For example, add other materials to the oil/water mix, try the activity with some of the dishes in a refrigerator, or place some dishes where they are exposed to direct sunlight.

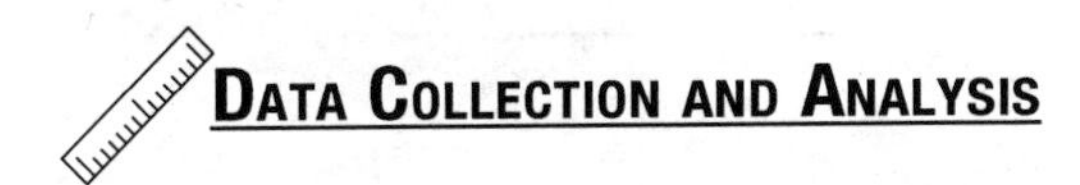

DATA COLLECTION AND ANALYSIS

SHELF DISHES

	Shelf 1	Shelf 2	Shelf 3	Shelf 4
Dish contains				
Day 1				
Day 2				
Day 3				
Day 4				
Day 5				
Day 6				
Day 7				

Name ______________________ Date ______________________

15. Oil Cleanup with Bacteria

STUDENT ACTIVITY PAGE

INCUBATOR DISHES

	Incubator 1	Incubator 2	Incubator 3	Incubator 4
Dish contains				
Day 1				
Day 2				
Day 3				
Day 4				
Day 5				
Day 6				
Day 7				

CONCLUDING QUESTIONS

1. Which container showed the greatest change in the nature of the oil?

____

____

____

____

2. Which container showed the least change in the nature of the oil?

____

____

____

____

Name ______________________ Date ______________________

15. Oil Cleanup with Bacteria

STUDENT ACTIVITY PAGE

3. How did the nature of the oil compare between the samples on the shelf and the samples in the incubator?

____

____

____

____

4. Based on the results of this activity, what conditions are best for cleaning an oil spill? (For example, should it be warm? Should it be cool? Should salt be added?)

____

____

____

____

FOLLOW-UP ACTIVITY

Find some examples of real-life bioremediation other than using biotechnology to clean up oil spills. Present your findings to the class in a format approved by your teacher.

16. Where Will the Money Go?

Teacher Resource Page

Instructional Objectives

Students will be able to:

- determine how to allocate limited resources
- define *conservation*
- understand conservation as a tool of biotechnology

National Science Education Standards Correlations

Grades 5–8

Content standard	Bullet number	Content description	Bullet number(s)
A	1	Abilities necessary to do scientific inquiry	3, 4, 7
A	2	Understandings about scientific inquiry	1, 4
C	3	Regulation and behavior	1, 4
C	5	Diversity and adaptations of organisms	1
E	1	Abilities of technological design	1, 2, 4
E	2	Understandings about science and technology	4, 6
F	5	Science and technology in society	3, 7

Grades 9–12

Content standard	Bullet number	Content description	Bullet number(s)
A	1	Abilities necessary to do scientific inquiry	1–6
A	2	Understandings about scientific inquiry	1, 3, 6
C	2	Molecular basis of heredity	1, 3
C	4	Interdependence of organisms	3, 5
E	1	Abilities of technological design	1–5
E	2	Understandings about science and technology	1–3
F	6	Science and technology in local, national, and global challenges	4
G	3	Historical perspectives	3

Vocabulary

- **conservation:** in general, the protection of existing habitats and species

Materials

- Internet access or other resources
- notebook

16. Where Will the Money Go?

Helpful Hints and Discussion

Time frame: one week for research
Structure: individuals or partners
Location: classroom, library

In this activity, students will be looking at the various ways that conservation is used to preserve or manipulate the population of a given species. They will research to determine how the management of a certain species has implications that are ethical in nature and can affect how science is used around these species. They will also see that some species are sought by tourists, and that means that maintaining them in their environment can have economic repercussions in an area as well. There are legal issues and time issues and, when it gets right down to it, there have to be people who are willing to take on a job.

Meeting the Needs of Diverse Learners

Students who need extra challenges should complete the Follow-Up Activity and the Extension Option. They should also be encouraged to work with others to help them better understand the way the various systems (ethical, scientific, economic, legal, etc.) are connected. Students who need extra help will benefit from a discussion of the way that endangered species can be used to provide positive economic effects in an area, such as when tourists hire guides to take them into the wilderness to see endangered species.

Scoring Rubric

Students meet the standard for this activity by:

- understanding the nature of limited funding for wildlife preservation
- answering the Concluding Questions
- correctly identifying ten endangered species
- rating the ten species based on research

Recommended Internet Sites

- **EndangeredSpecie.com**
  www.endangeredspecie.com
- **The Endangered Species Act—Endangered Species Act Spending**
  www.libraryindex.com/pages/3033/Endangered-Species-Act-ENDANGERED-SPECIES-ACT-SPENDING.html
- **U.S. Fish & Wildlife Service Endangered Species Program**
  www.fws.gov/endangered

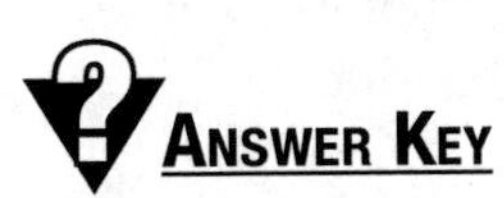

Answer Key

1–4. Answers will vary depending on the species students rated.

Name ____________________ Date ____________________

16. Where Will the Money Go?

STUDENT ACTIVITY PAGE

OBJECTIVE

To understand how wildlife programs and conservation programs decide to spend their limited funding

BEFORE YOU BEGIN

Biotechnology often means manipulating organisms to help people. But sometimes we manipulate organisms for their own good. There are numerous **conservation** programs that try to protect the natural environment for the species that live there. There are also programs that seek to reduce the damage already done by humans. Such work often requires resources and especially money. In this activity, you will get the chance to determine where the money is going to go.

MATERIALS

- Internet access or other resources
- notebook

PROCEDURE

1. Using the Internet or print resources, find ten endangered species in the United States. Learn about each one's habitat and how it is viewed by people who live in the same area.
2. Record the name of each organism in the data table in the Data Collection and Analysis section.
3. Rate each animal on a scale of 1 to 10 in each of the following categories, with 1 being least important for funding and 10 being most important for funding.
   - **Ethical concerns:** This category represents the ethics of allowing this particular species to go extinct. You may feel it would be more important to protect an endangered gorilla than an endangered slug. The choice might be unpleasant, but it is a choice that conservationists are sometimes faced with.
   - **Positive economic concerns:** Will this organism bring money into the local economy?
   - **Negative economic concerns:** Will money be lost from the local economy by protecting this organism? (Perhaps a business will have to close to protect the organism's habitat.)
   - **Legal concerns:** Will helping this organism lead to a long, expensive legal battle? (Perhaps a local business will sue if it will lose business because a protected species is where the business carries out many of its functions.)
   - **Scientific concerns:** What value does this animal have to science? To biotechnology? To scientists? To the environment?
   - **Overall concerns:** Your gut feeling about the importance of protecting this species based on what you know about it.
4. Add the numbers in each row and then list the species in the order of highest total points to lowest total points. The highest point total represents the species that will get the most funding.

Name ______________________ Date ______________________

16. Where Will the Money Go?

STUDENT ACTIVITY PAGE

EXTENSION OPTION

Trade your list of ten animals with another student, then rate that person's animals according to your own standards. How do the final ratings compare?

DATA COLLECTION AND ANALYSIS

Animal	Ethical concerns	Positive economic concerns	Negative economic concerns	Legal concerns	Scientific concerns	Overall concerns	Totals
1							
2							
3							
4							
5							
6							
7							
8							
9							
10							

CONCLUDING QUESTIONS

1. Which species will get the most funding?

____

____

____

____

16. Where Will the Money Go?

2. Which species will get the least funding?

____

____

____

____

3. How does the "attractiveness" of the highest-funded species compare with that of the lowest-funded species? For example, is the highest-funded species cute and furry and the lowest-funded species wet and slimy?

____

____

____

____

4. Think of one other concern that might be important to you. Rate the species on this concern and recalculate the totals. How did it change the order, if at all?

____

____

____

____

Follow-Up Activity

Research to determine what percent of the United States' federal budget is used to protect endangered species. Present your findings to the class in a format approved by your teacher.

17. Enzymes

TEACHER RESOURCE PAGE

INSTRUCTIONAL OBJECTIVES

Students will be able to:

- describe the behavior of organic materials treated with enzymes
- understand how biotechnology can use enzymes as tools

NATIONAL SCIENCE EDUCATION STANDARDS CORRELATIONS

GRADES 5–8

Content standard	Bullet number	Content description	Bullet number(s)
A	1	Abilities necessary to do scientific inquiry	3, 4, 7
A	2	Understandings about scientific inquiry	1, 4
C	1	Structure and function in living systems	1, 2
C	2	Reproduction and heredity	3, 4
E	2	Understandings about science and technology	4, 6
F	5	Science and technology in society	3, 7

GRADES 9–12

Content standard	Bullet number	Content description	Bullet number(s)
A	1	Abilities necessary to do scientific inquiry	1–6
A	2	Understandings about scientific inquiry	1, 3, 6
B	3	Chemical reactions	1, 5
C	1	The cell	1–4
C	2	Molecular basis of heredity	1, 3
E	1	Abilities of technological design	1–5
E	2	Understandings about science and technology	1–3
G	3	Historical perspectives	3

VOCABULARY

- **enzymes:** biological molecules that act as catalysts for chemical reactions

17. Enzymes

TEACHER RESOURCE PAGE

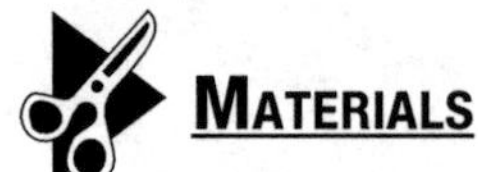

MATERIALS

- RID-X Septic System Treatment (liquid)
- bottled water
- bread (cubed, approximately 2 cm × 2 cm × 2 cm)
- banana (sliced into 3-cm pieces)
- raw liver (cut into pieces about the size of half a thumb)
- apple (cubed, approximately 2 cm × 2 cm × 2 cm)
- plastic wrap
- 50-ml beaker
- 400-ml beakers (or larger); eight per team
- rubber bands
- goggles
- lab aprons
- gloves
- marking pencil or tape for labeling beakers

HELPFUL HINTS AND DISCUSSION

Time frame: three days
Structure: partners
Location: classroom

In this activity, students will witness the power of bacteria and enzymes to digest a variety of food products. They will have two sets of food items, one which will be treated with RID-X and one that will be left to degrade without any assistance. It is important that you use bottled water and not tap water, because the chlorine in tap water will interfere with the reproduction of the microorganisms in the RID-X. To save class time, cut the liver, apples, bread, and bananas into pieces of the indicated sizes ahead of time, or have trustworthy students cut them up. Bananas and apples will brown upon exposure to air, but this should not affect results.

MEETING THE NEEDS OF DIVERSE LEARNERS

Students who need extra challenges should complete the Follow-Up Activity and the Extension Option. These students should also be encouraged to work with struggling students while they make their observations during each class period.

SCORING RUBRIC

Students meet the standard for this activity by:

- safely handling lab equipment
- recording observations accurately and thoroughly
- using clear and descriptive language observations
- answering the Concluding Questions

17. Enzymes

RECOMMENDED INTERNET SITES

- **Enzyme Stuff—How Enzymes Work**
  www.enzymestuff.com/basicswhichenzyme.htm#1
- **Procter&Gamble—What Are Enzymes and Why Do We Use Them in Laundry Detergents?**
  www.scienceinthebox.com/en_UK/safety/whatareenzymes_en.html
- **RID-X Septic System Treatment—Frequently Asked Questions**
  www.rid-x.com/faq.shtml

ANSWER KEY

1. The bread and banana generally show the most change. Results will vary.
2. The apple and the liver generally show the least change. Results will vary.
3. The samples with the RID-X should degrade faster.
4. Yes, because a product such as RID-X could help break down organic materials in both arenas.

Name ______________________________ Date ______________________________

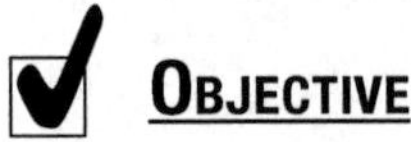

17. Enzymes

STUDENT ACTIVITY PAGE

OBJECTIVE

To see how enzymes increase the speed at which biological compounds break down

BEFORE YOU BEGIN

Enzymes are biological molecules that act as catalysts for chemical reactions. These include reactions that occur inside living organisms. Biotechnologists have found ways to apply the use of enzymes to the production of paper, cheese, and beer. Enzymes are also used to clean contact lenses and can be found in detergents and sewage-treatment products. They are also used in many other industries. Of course, they are found performing many tasks throughout the human body. One product that contains a large amount of bacteria and enzymes is RID-X. RID-X is designed to help microorganisms as they digest the household waste found in a septic system. It is not used in municipal water-treatment systems. This activity will give you the chance to see the kind of power these microorganisms and enzymes have.

MATERIALS

- RID-X Septic System Treatment (liquid)
- bottled water
- bread (cubed, approximately 2 cm × 2 cm × 2 cm)
- banana (sliced into 3-cm pieces)
- raw liver (cut into pieces about the size of half a thumb)
- apple (cubed, approximately 2 cm × 2 cm × 2 cm)
- plastic wrap
- 50-ml beaker
- 400-ml beakers (or larger); eight per team
- rubber bands
- goggles
- lab apron
- gloves
- marking pencil or tape for labeling beakers

PROCEDURE

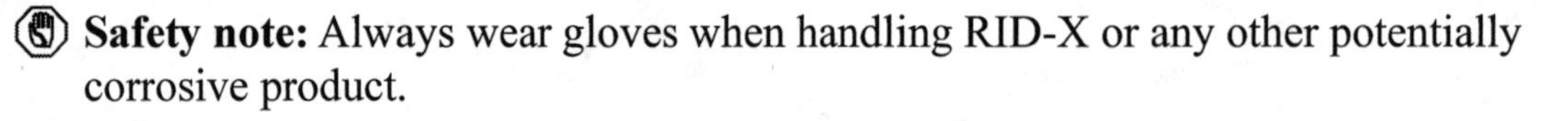

Safety note: Always wear gloves when handling RID-X or any other potentially corrosive product.

1. Label your beakers 1 through 8.
2. Add 200 ml of water to each beaker.
3. Add 20 ml of RID-X liquid to beakers 1, 3, 5, and 7 only. DO NOT add RID-X to the other beakers.
4. Add a cube of bread approximately 2 cm × 2 cm × 2 cm to beakers 1 and 2.
5. Add a 3-cm-long slice of banana to beakers 3 and 4.
6. Add a piece of liver about the size of half your thumb to beakers 5 and 6.

Name ______________________ Date ______________________

17. Enzymes

STUDENT ACTIVITY PAGE

7. Add a cube of apple approximately 2 cm × 2 cm × 2 cm to beakers 7 and 8.
8. Cover the top of the beakers with plastic wrap. Put a rubber band around the top of each beaker to help hold the plastic in place.
9. Record observations for all eight beakers in your data table under Day 1.
10. Repeat observations on Days 2 and 3, and then give the beakers to your teacher for disposal. If a fume hood is available, try using a stirring rod to poke the samples and move them around to get some idea of how each sample's texture has changed. **Do not touch or smell the materials in the beakers. The material is not safe to eat.**

EXTENSION OPTION

Repeat this experiment but try leaving the beakers uncovered and in a fume hood. What differences, if any, do you see?

DATA COLLECTION AND ANALYSIS

Beaker	Contents	Day 1	Day 2	Day 3
1	RID-X, bread, water			
2	Bread, water			
3	RID-X, banana, water			
4	Banana, water			
5	RID-X, liver, water			
6	Liver, water			
7	RID-X, apple, water			
8	Apple, water			

Name ______________________ Date ______________________

17. Enzymes

STUDENT ACTIVITY PAGE

Concluding Questions

1. Which substance showed the greatest change across the three days?

____

____

____

____

2. Which substance showed the least change across the three days?

____

____

____

____

3. How did the samples with RID-X look compared to the samples without RID-X?

____

____

____

____

4. Is it likely that something such as RID-X would be beneficial in composting or bioremediation? Explain.

____

____

____

____

Follow-Up Activity

Find some other products that contain enzymes. What is each product used for? Is it clearly a biotechnology application? Present your findings to the class in a format approved by your teacher.

18. Forensics

INSTRUCTIONAL OBJECTIVES

Students will be able to:

- analyze a sample to see if it contains blood
- connect forensic science with biotechnology
- collect simulated blood samples
- draw conclusions about the nature of the sample they have collected

NATIONAL SCIENCE EDUCATION STANDARDS CORRELATIONS

GRADES 5–8

Content standard	Bullet number	Content description	Bullet number(s)
A	1	Abilities necessary to do scientific inquiry	3, 4, 7
A	2	Understandings about scientific inquiry	1, 4
C	1	Structure and function in living systems	1, 2
C	2	Reproduction and heredity	3, 4
E	2	Understandings about science and technology	4, 6
F	5	Science and technology in society	3, 7

GRADES 9–12

Content standard	Bullet number	Content description	Bullet number(s)
A	1	Abilities necessary to do scientific inquiry	1–6
A	2	Understandings about scientific inquiry	1, 3, 6
B	3	Chemical reactions	1
C	1	The cell	1–4
E	1	Abilities of technological design	1–5
E	2	Understandings about science and technology	1–3
G	3	Historical perspectives	3

VOCABULARY

- **ABO system:** one of several systems for determining blood types in humans
- **blood:** the fluid that circulates in the principal vascular system, consisting of red blood cells, white blood cells, and platelets suspended in plasma; contains information, such as DNA, that is specific to a particular individual
- **blood typing:** a way of classifying blood based on the various chemicals it may contain
- **chemiluminescence:** a chemical reaction that gives off light
- **Fluorescin:** a chemical used to reveal the presence of blood evidence
- **luminol:** a chemical used to reveal the presence of blood evidence

18. Forensics

Teacher Resource Page

Materials

- artificial blood
- sterile swabs
- distilled water
- rinse bottles
- small beakers, 50 or 100 ml
- luminol
- potassium hydroxide
- hydrogen peroxide (3% over-the-counter concentration)
- coffee
- cola
- cranberry juice
- red food dye
- old clothing or rags
- gloves (latex and nitrile)
- goggles
- color chart
- lab aprons

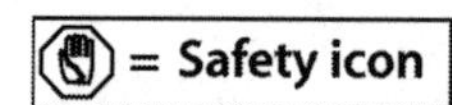

Helpful Hints and Discussion

Time frame: one class period
Structure: individuals or groups
Location: classroom and an area that can be made dark, such as a closet or storeroom

Most school districts do not allow students to handle real blood samples of any kind. Be sure to use artificial blood. If your school allows blood testing, be sure to request a copy of the guidelines and apply them to this activity.

The Web sites below both sell artificial blood that contains no blood products and that reacts with luminol:

- www.evidentcrimescene.com
- www.crimescene.com

You will need to pre-stain some old clothing, rags, or surfaces with artificial blood, cola, cranberry juice, red food dye, and coffee and allow them time to dry before the lab.

Some students may be allergic to latex gloves, so be sure to have nitrile gloves available.

To make a stock luminol solution, mix 8 g of luminol with 60 g of potassium hydroxide and 1000 ml of distilled water.

Safety note: Special care should be used when handling potassium hydroxide as it is corrosive. The finished luminol solution will need to be handled with care as well as it will contain the potassium hydroxide.

18. Forensics

TEACHER RESOURCE PAGE

MEETING THE NEEDS OF DIVERSE LEARNERS

Students who need extra challenges should complete the Follow-Up Activity and the Extension Option. These students should also be encouraged to explore other aspects of biotechnology used in forensics. Students who need extra help will benefit from a careful walk-through of the procedure.

SCORING RUBRIC

Students meet the standard for this activity by:

- using correct procedure for collecting and testing apparent blood samples
- recording data that indicates or does not indicate the presence of blood
- using scientific terms such as *luminol* and *chemiluminescence* correctly
- answering the Concluding Questions accurately

RECOMMENDED INTERNET SITES

- **Collection and Preservation of Blood Evidence from Crime Scenes**
  www.crime-scene-investigator.net/blood.html
- **Suite101.com—Luminol: Chemiluminescent Blood Detector**
  http://crime-scene-processing.suite101.com/article.cfm/chemiluminescent_luminol

ANSWER KEY

1. Most students have seen dried blood and are pretty good at recognizing it when they see it. Depending on the quality of the artificial blood, students may think that it is not blood.
2. Students frequently notice the smell of coffee. Answers will vary.
3. The luminol test requires a dark workplace and is difficult to do at a crime scene. This is why many samples are not tested until they are back at a lab.
4. There are many different kinds of forensic scientists. One who exclusively studies DNA and related biomolecules might be thought of as a biotechnologist, while a forensic scientist who studies skid-mark length and glass fragments might not.

Name ____________________ Date ____________________

18. Forensics

STUDENT ACTIVITY PAGE

OBJECTIVE

To determine whether or not a collected sample is blood

BEFORE YOU BEGIN

The human body contains 5 to 6 liters of **blood**, and that blood is made of many biological components. Biotechnology has provided a large number of tools for examining blood. Forensic scientists can determine what chemicals are in a sample and can further identify individual-specific compounds such as DNA in blood. Blood can be found at a crime scene in a number of ways. There is the obvious appearance of blood as drops or pools that can be seen with the naked eye. Sometimes chemical agents are required to reveal the presence of blood evidence, especially if someone has tried to cover it up. You can spray **luminol** solution on a location where you suspect blood was spilled and then cleaned. If blood residue is present, the blood and luminol will react with each other and glow light blue in the dark. This chemical reaction that produces light is called **chemiluminescence. Fluorescin** is another chemical that is far more sensitive than luminol. However, it has the disadvantage that it can only indicate the presence of blood when the sample is looked at under ultraviolet light.

More of the story of your blood can be found through blood typing. There are a number of ways that blood is classified. The **ABO system** views human blood as having four basic blood types: A, B, O, and AB. There are other systems that recognize far more individual parts of blood that would account for almost 600 testable components. While **blood typing** is fast being replaced with DNA matching, blood-typing tests are relatively inexpensive and quick. They can be used to determine if two blood samples are a general match or not.

MATERIALS

- unknown samples
- sterile swabs
- distilled water
- rinse bottle
- small beakers, 50 or 100 ml
- (S) luminol solution
- hydrogen peroxide
- gloves
- goggles
- color chart
- lab apron

(S) = Safety icon

Name ________________________________ Date ________________________

18. Forensics

STUDENT ACTIVITY PAGE

PROCEDURE

> **Safety note:** Wear your gloves and goggles while you handle the unknown samples and testing chemicals, especially the luminol solution. This protects both you and the sample.

1. Look at the unknown samples provided by your teacher. In the Data Collection and Analysis section, write a description of each stain in your data table before you disturb the stain. Indicate the color and the size. Do any of them look like blood?
2. Find the closest similar color on the color chart for the first stain. Record this color in the data table.
3. Wet your sterile swab with distilled water from the rinse bottle.
4. Wipe the swab through the stain.
5. Mix 10 ml of luminol solution and 10 ml of hydrogen peroxide in a small beaker.

> **Safety note:** Be careful handling the luminol solution as it contains corrosive chemicals. If you get it on your skin, tell your teacher and wash the chemicals off your skin immediately.

6. Take your sample and swab to the dark testing area and dip the swab into the solution. A blue glow indicates the presence of blood.
7. Record whether there was a positive or negative reaction for the sample.
8. Repeat steps 2 through 7 for the other samples. **Be sure to use a new swab and new luminol and hydrogen peroxide solution each time. Use a clean beaker for each sample. Reusing the swab, the luminol solution, or a dirty beaker will contaminate your results.**

EXTENSION OPTION

You have more than the sense of sight to rely on when collecting data. With your teacher's permission, smell each of the samples and see if you can identify them by smell. Remember, this is not always a safe way to test a sample, but some materials will have an odor you can smell whether you want to or not.

Name ______________________ Date ______________________

18. Forensics

STUDENT ACTIVITY PAGE

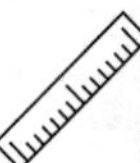

DATA COLLECTION AND ANALYSIS

Sample number	Where collected	Description (color, size, shape)	Luminol test (positive or negative)

CONCLUDING QUESTIONS

1. Which of the samples did you think were blood before they were tested?

____

____

____

____

2. Did you notice any odors from the samples before you tested them? If so, what did they smell like?

____

____

____

____

18. Forensics

3. What part of this test, if any, would be difficult to carry out during the daytime at an outside crime scene?

____

____

____

____

4. Do you think a forensic scientist would consider himself or herself to be a biotechnologist? Why or why not?

____

____

____

____

FOLLOW-UP ACTIVITY

Find out at least five other ways that biotechnology is used in forensics. Present your findings to the class in a format approved by your teacher.

19. Extracting DNA

TEACHER RESOURCE PAGE

Instructional Objectives

Students will be able to:

- extract DNA from a fruit sample
- understand DNA extraction as a tool of biotechnology
- explain the importance of the various steps of the procedure

National Science Education Standards Correlations

Grades 5–8

Content standard	Bullet number	Content description	Bullet number(s)
A	1	Abilities necessary to do scientific inquiry	3, 4, 7
A	2	Understandings about scientific inquiry	1, 4
C	1	Structure and function in living systems	1, 2
C	2	Reproduction and heredity	3, 4
E	2	Understandings about science and technology	4, 6
F	5	Science and technology in society	3, 7

Grades 9–12

Content standard	Bullet number	Content description	Bullet number(s)
A	1	Abilities necessary to do scientific inquiry	1–6
A	2	Understandings about scientific inquiry	1, 3, 6
B	3	Chemical reactions	1, 5
C	1	The cell	1–4
C	2	Molecular basis of heredity	1, 3
C	4	Interdependence of organisms	3, 5
E	1	Abilities of technological design	1–5
E	2	Understandings about science and technology	1–3
G	3	Historical perspectives	3

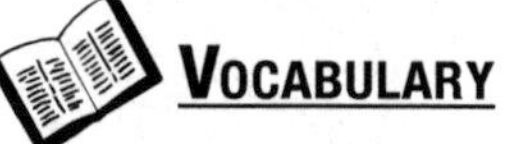

Vocabulary

- **DNA:** deoxyribonucleic acid; a nucleic acid that carries the genetic information in the cell

19. Extracting DNA

TEACHER RESOURCE PAGE

MATERIALS

- kiwifruit or strawberry
- shampoo (without conditioner, not baby shampoo)
- salt
- resealable plastic bags
- ice bath
- cheesecloth
- funnel
- beaker
- ring stand
- ring
- distilled water
- 600-ml beaker
- test tube
- test-tube holder
- 95% ethanol (refrigerated)
- wooden splint
- goggles
- lab aprons
- gloves

HELPFUL HINTS AND DISCUSSION

Time frame: 60 minutes
Structure: individuals or partners
Location: classroom

This activity is a brute-force, chemistry-based extraction of DNA. The final product is not the purified version used in advanced biotechnology applications, but it is relatively pure DNA. You will need to prepare the shampoo/salt solution by adding 50 ml of shampoo (without conditioner and not baby shampoo) to 10 g of salt in a 600-ml beaker. Mix the shampoo and salt slowly and carefully to reduce the buildup of foam, and then add distilled water to bring the total volume up to 500 ml.

The students may or may not be aware what each step accomplishes in the activity, so a summary is listed below. Review these with students as they go through the steps to ensure that they understand.

- Mashing the kiwifruit helps break down the cell walls, as does the shampoo.
- The addition of salt allows unwanted proteins and carbohydrates to precipitate out.
- The cold water bath slows the enzyme activity that would normally destroy the DNA after the cell membranes are broken down.
- The cheesecloth helps catch the large pieces of fruit, as well as the precipitated proteins and carbohydrates.
- Adding cold ethanol increases the rate of precipitation for the DNA.

19. Extracting DNA

TEACHER RESOURCE PAGE

MEETING THE NEEDS OF DIVERSE LEARNERS

Students who need extra challenges should be encouraged to explain what each step of the procedure is accomplishing before starting the activity. Students who need extra help will benefit from a careful walk-through of the procedure, including setting up the equipment.

SCORING RUBRIC

Students meet the standard for this activity by:

- successfully collecting a sample of DNA from the fruit
- correctly following the procedure
- safely handling lab equipment and chemicals
- answering the Concluding Questions

RECOMMENDED INTERNET SITES

- **Carnegie Academy for Science Education—Berry Full of DNA (lesson plan containing additional student response questions)**
  http://carnegieinstitution.org/first_light_case/horn/DNA/BERRYteacDNA
- **Microbial Life Educational Resources—DNA Extraction**
  http://serc.carleton.edu/microbelife/research_methods/genomics/dnaext.html
- **The University of Utah Genetic Science Learning Center—How to Extract DNA from Anything Living**
  http://learn.genetics.utah.edu/content/labs/extraction/howto/

ANSWER KEY

1. Mashing the fruit helps break the cell membranes.
2. The shampoo helps break the cell membranes, and the salt allows proteins and carbohydrates we do not want to precipitate out.
3. The proteins and carbohydrates are filtered out by the cheesecloth. Some other waste parts of the cell are also filtered out.
4. The cold ethanol helps increase the amount of DNA that precipitates out.

Name ______________________________ Date ______________________________

19. Extracting DNA

Student Activity Page

Objective

To extract a DNA sample from fruit

Before You Begin

The first step of many advanced procedures in biotechnology is the extraction of **DNA** from an organism. DNA is found in the cells of all living things. Once samples of the DNA are removed, biotechnologists can do a variety of things with them. They can modify them, purify the sample, test the sample for evidence of a genetic illness, or use them to identify the organism from which the sample came. The more you learn about biotechnology, the more places you will see DNA used as the starting point. In this activity, you will have the chance to extract DNA from fruit and see what it looks like in a collected form.

Materials

- kiwifruit or strawberry
- shampoo/salt solution
- resealable plastic bags
- ice bath
- cheesecloth
- funnel
- beaker
- ring stand
- ring
- test tube
- test-tube holder
- 95% ethanol (refrigerated)
- wooden splint
- goggles
- lab apron
- gloves

Procedure

1. Place one quarter of a kiwifruit (peeled) or one half of a large strawberry in the resealable plastic bag.
2. Add 20 ml of the shampoo/salt solution.
3. Close the bag while getting as much of the air out as possible.
4. Mash the fruit gently with your fingers by squishing the bag. Do this for 3 to 5 minutes.
5. Choose the bag in the ice bath for 1 minute, then mush for 1 minute. Do this a total of three times.
6. Place the cheesecloth over the top of the funnel. Place the covered funnel over a beaker.
7. Slowly dump the contents of the bag into the funnel. Collect the fruit solution in the beaker.
8. Put 5 ml of the solution in a test tube.
9. DO NOT MOVE the test tube. SLOWLY add 5 ml of refrigerated ethanol to the test tube and let it run down the inside of the tube. DO NOT dump it quickly.

Name ______________________________ Date ______________________________

19. Extracting DNA

STUDENT ACTIVITY PAGE

10. A substance should form at the top of the solution. Collect it using a wooden splint. This material is the fruit's DNA.
11. Record your observations about the appearance of the DNA in the Data Collection and Analysis section.

EXTENSION OPTION

Try this extraction with other fruits and vegetables. Compare the appearances of the final products from each attempt.

DATA COLLECTION AND ANALYSIS

DNA observations:

____

____

CONCLUDING QUESTIONS

1. Why did you mash the fruit with your hands?

____

____

____

____

2. What task was performed by the shampoo/salt solution?

____

____

____

____

3. What material or materials were filtered out by the cheesecloth?

____

____

____

____

19. Extracting DNA

4. Why was cold ethanol added in step 9?

____

____

____

____

 **FOLLOW-UP ACTIVITY**

Research to discover five biotechnology applications that start with the extraction of DNA from an organism. Present your findings to the class in a format approved by your teacher.

20. Biotechnology Careers

TEACHER RESOURCE PAGE

INSTRUCTIONAL OBJECTIVES

Students will be able to:

- describe the educational background needed for each of the five jobs
- describe five careers in biotechnology

NATIONAL SCIENCE EDUCATION STANDARDS CORRELATIONS

GRADES 5–8

Content standard	Bullet number	Content description	Bullet number(s)
A	1	Abilities necessary to do scientific inquiry	3, 4, 7
A	2	Understandings about scientific inquiry	1, 4
F	5	Science and technology in society	3, 7

GRADES 9–12

Content standard	Bullet number	Content description	Bullet number(s)
A	1	Abilities necessary to do scientific inquiry	1–6
A	2	Understandings about scientific inquiry	1, 3, 6
C	1	The cell	1–4
E	1	Abilities of technological design	1–5
E	2	Understandings about science and technology	1–3
F	6	Science and technology in local, national, and global challenges	4
G	3	Historical perspectives	3

VOCABULARY

- **biotechnology:** the use or modification of organisms for human purposes

MATERIALS

- Internet access or other resources
- notebook

20. Biotechnology Careers

TEACHER RESOURCE PAGE

HELPFUL HINTS AND DISCUSSION

Time frame: one week for research

Structure: individuals

Location: classroom, library, guidance office, home

There are so many jobs in biotechnology, it is practically impossible to know what they all entail. This activity is designed to allow students to collect information about five careers in biotechnology and the prerequisite education needed to be considered for those jobs. They may find that your school's guidance office has access to material that would clarify the educational background needed for each job, or at least the major to concentrate on in their post-secondary education. Students will need to choose a favorite possible career to share with their peers in a brief presentation at the end of the activity.

MEETING THE NEEDS OF DIVERSE LEARNERS

Students who need extra challenges should complete the Follow-Up Activity and the Extension Options, especially the option concerning research into biotechnology scholarships, as the information gleaned would benefit the entire class. These students should also be encouraged to look into more than five careers, especially if they are interested a biotechnology job. Students who need extra help will benefit from assistance narrowing the list of possible jobs to research.

SCORING RUBRIC

Students meet the standard for this activity by:

- correctly identifying educational background needed for each career
- presenting information to classmates about chosen career
- collecting a list of 20 biotechnology careers
- summarizing five careers
- answering the Concluding Questions

RECOMMENDED INTERNET SITES

- **Access Excellence @ The National Health Museum Resource Center—Career Center**
  www.accessexcellence.org/RC/CC
- **Biotechnology Information Series—Careers in Biotechnology**
  www.biotech.iastate.edu/biotech_info_series/bio2.html
- **Biotechnology Institute—Careers in Biotechnology**
  www.biotechinstitute.org/careers
- **Career Voyages—Biotechnology Jobs in the News**
  www.careervoyages.gov/biotechnology-main.cfm

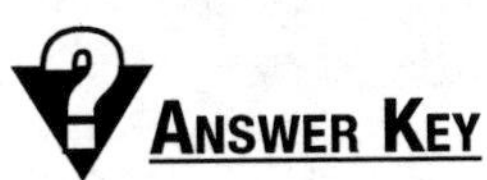

ANSWER KEY

1–4. Answers will vary.

Name ____________________ Date ____________________

20. Biotechnology Careers

STUDENT ACTIVITY PAGE

OBJECTIVE

To become familiar with some careers in biotechnology and the education needed to do those jobs

BEFORE YOU BEGIN

Biotechnology jobs can be found in almost any industry. Some jobs are entry-level and require very little education beyond high school. Others are very advanced, requiring many years of college and sometimes degrees in more than one field. It is important to remember that someone making cheese and someone developing a form of gene therapy are both engaged in biotechnology. With that in mind, this activity will help you learn about a number of biotechnology jobs. Then you will narrow your search down to five careers that you find interesting.

MATERIALS

- Internet access or other resources
- notebook

PROCEDURE

1. Using the Internet or print resources, make a list of 20 jobs that could be considered careers in biotechnology.
2. Choose five from your list. They should be careers you would like to work at, or should at least be careers that you find interesting. Have your teacher approve your list.
3. Use the data table in the Data Collection and Analysis section to collect as much information as you can about each career. Find the names of companies that employ people in that career. Determine which, if any, post-high-school degrees you would need or training you would need to be working at that career. What kind of salary range might you expect to earn in each career?
4. Write a summary of approximately one page per career to turn in to your teacher.
5. Choose your favorite career from your list of five. Prepare a brief presentation (5 minutes or less) to share the details of that career with your classmates. Your teacher may choose to have you present more than one career if time allows.

EXTENSION OPTIONS

1. Choose one career that did not seem as interesting to you and perform the same research as you did for the five you were interested in. Did the career seem more interesting after you had learned more about it?
2. Research scholarships available to students who pursue careers in biotechnology. Present your findings to the class in a format approved by your teacher.

Name ______________________ Date ______________

20. Biotechnology Careers

STUDENT ACTIVITY PAGE

Data Collection and Analysis

Top Five Careers

Career	Potential employers	Education/ training needed	Salary range (starting–top)

Concluding Questions

1. Which career seemed most interesting to you? Why?

____

____

____

____

Name ______________________ Date ______________

20. Biotechnology Careers

STUDENT ACTIVITY PAGE

2. Did any of the careers seem more interesting after you learned more about them? Explain.

___

___

___

___

3. Which, if any, of the careers would you be interested in following? Why?

___

___

___

___

4. Were there any careers you think that you would actively dislike? Explain.

___

___

___

___

FOLLOW-UP ACTIVITY

Locate someone working in a biotechnology field and ask him or her to share the things he or she likes and does not like about that job. Present your findings to the class in a format approved by your teacher.

Glossary

ABO system: one of several systems for determining blood types in humans

antibiotic: a substance that kills bacteria or at least hinders the growth of bacteria

bacillus thuringiensis (Bt): a kind of bacteria found in soil that can be used as a pesticide

bacteria: single-celled microorganisms

bioremediation: the process of breaking down or removing toxins from the environment using living organisms or components of living organisms, such as plants, enzymes, fungi, and any of the many kinds of microorganisms

biotechnology: the use or modification of organisms for human purposes

Biuret solution: a solution used to identify the likely presence of proteins in a sample

blood: the fluid that circulates in the principal vascular system, consisting of red blood cells, white blood cells, and platelets suspended in plasma; contains information, such as DNA, that is specific to a particular individual

blood typing: a way of classifying blood based on the various chemicals it may contain

by-products: secondary products produced during or after the production of a main product, such as eggs from chickens, assuming chickens are the main product

chemiluminescence: a chemical reaction that gives off light

cloning: the process of reproducing organisms so that they have a genetic makeup that is identical to that of the original organism

conservation: in general, the protection of existing habitats and species

DNA: deoxyribonucleic acid; a nucleic acid that carries the genetic information in the cell

electrophoresis: the motion of electrically charged particles in a material that is permeated by an electric field

enzymes: biological molecules that act as catalysts for chemical reactions

ethical: pertaining to behavior as decided by a group or society

Fluorescin: a chemical used to reveal the presence of blood evidence

fungi: organisms of the taxonomic kingdom Fungi that include mushrooms, molds, toadstools, and yeasts

gel electrophoresis: a process for separating large organic molecules into smaller component molecules using an electric field directed through a gel

genetic engineering: a process of manipulating an organism's genes

genetically modified food: food from an organism that has had some part of its genetic makeup altered by genetic engineering

genetically modified organism (GMO): an organism that has had some part of its genetic makeup altered by genetic engineering

insecticide: a toxin used to kill insects

invasive species: species that have moved into an environment where they are not typically found, generally with negative effects

kimchi: a popular Korean dish that generally contains pickled cabbage

lactobacilli: a kind of bacteria that usually convert lactose into lactic acid

luminol: a chemical used to reveal the presence of blood evidence

microbes: generally organisms that are microscopic in size

moral: relating to principles of right and wrong

non-native species: species that are not normally found in a given environment

pickling: a form of lactic acid fermentation usually used for increasing the storage time for vegetables

Glossary

proteins: organic compounds made of amino acids that are found in all living organisms

resistant: a characteristic of an organism that, generally through selection, has become immune to the effects of a substance

toxins: poisons produced by or made from living organisms

vegetative propagation: the process of cloning plants from cuttings

yeast: generally a unicellular organism from the Fungi kingdom

yogurt: a dairy product that is made from milk though lactic acid fermentation

extending and enhancing learning

Let's stay in touch!

Thank you for purchasing these Walch Education materials. Now, we'd like to support you in your role as an educator. **Register now** and we'll provide you with updates on related publications, online resources, and more. You can register online at www.walch.com/newsletter, or fill out this form and fax or mail it to us.

Name ______________________________ Date ______________

School name ______________________________

School address ______________________________

City ______________________ State ______________ Zip ________

Phone number (home) ____________________ (school) ____________________

E-mail ______________________________

Grade level(s) taught ___________ Subject area(s) ______________________

Where did you purchase this publication? ______________________

When do you primarily purchase supplemental materials? ______________________

What moneys were used to purchase this publication?

[ ] School supplemental budget

[ ] Federal/state funding

[ ] Personal

[ ] Please sign me up for Walch Education's free quarterly e-newsletter, *Education Connection.*

[ ] Please notify me regarding free *Teachable Moments* downloads.

[ ] Yes, you may use my comments in upcoming communications.

COMMENTS ______________________________

____

____

____

Please FAX this completed form to 888-991-5755, or mail it to:

Customer Service, Walch Education, 40 Walch Drive, Portland, ME 04103